BELLWORK®

A DAILY PRACTICE PROGRAM

Math Level 4

kohlton

BELLWORK Enterprises, Inc.
Stanton, CA

Author

Charles W. De Pue

Co-Authors:

Margaret Kinney

Kent A. De Pue

Illustrator

José L. de la Rosa

BELLWORK
Stanton, CA
(800) 782-8869

Printed in the U.S.A.
ISBN 0-934475-86-5

Name ______________________________

1

$$\begin{array}{r} 29 \\ -\ 20 \\ \hline \end{array}$$

Ⓐ 9

Ⓑ 8

Ⓒ 10

Ⓓ 49

2

$$\begin{array}{r} 31 \\ +\ 25 \\ \hline \end{array}$$

Ⓕ 54

Ⓖ 14

Ⓗ 66

Ⓙ 56

3 **Jim's cat had 7 kittens. He found homes for 4 of them. How many kittens does he have left?**

Ⓐ 11 kittens

Ⓒ 3 kittens

Ⓑ 2 kittens

Ⓓ 4 kittens

4 **Which figure below is a square?**

Ⓕ △

Ⓗ □

Ⓖ ▭

Ⓙ ○

Name ______________________________

1 **Look at the coins below.**

How much money does this represent?

Ⓐ 87¢
Ⓑ 77¢
Ⓒ 57¢
Ⓓ 97¢

2 **Which list has the numbers in order from greatest value to least value?**

Ⓕ 1312, 806, 904, 1027
Ⓖ 806, 904, 1027, 1312
Ⓗ 1312, 1027, 806, 904
Ⓙ 1312, 1027, 904, 806

3 **Imagine that it is snowing outside. What is most likely the temperature today?**

Ⓐ 30°F
Ⓑ 90°F
Ⓒ 60°F
Ⓓ 110°F

Name ______________________________

1 **Kent and Robyn went fishing. Kent caught 4 fish. What else do you need to know to find the number of fish they caught all together?**

Ⓐ How big was each fish?

Ⓑ How long did they fish?

Ⓒ Why did they go fishing?

Ⓓ How many fish Robyn caught?

2 **Which list has the numbers in order from least value to greatest value?**

Ⓕ 8841, 8861, 8792, 7993

Ⓖ 8792, 7993, 8841, 8861

Ⓗ 7993, 8792, 8841, 8861

Ⓙ 7993, 8792, 8861, 8841

3 **Look at the coins below.**

How much money does this represent?

Ⓐ 91¢

Ⓑ 96¢

Ⓒ 86¢

Ⓓ 81¢

Name ____________________

1

$2 \times 2 = \square$

Ⓐ 2 Ⓒ 5

Ⓑ 22 Ⓓ 4

3

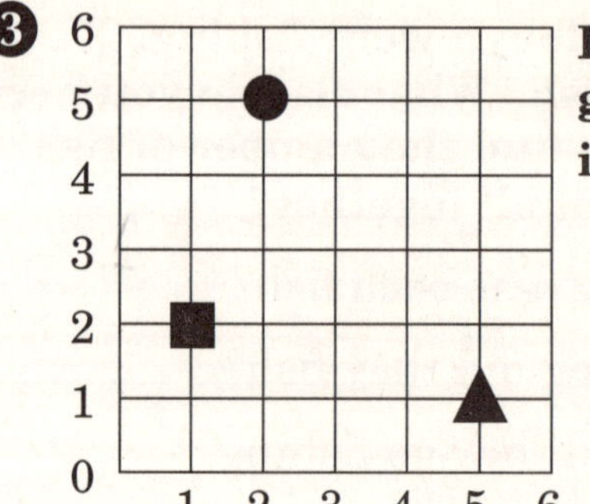

Look at the graph. Where is the ● ?

Ⓐ (3,4)

Ⓑ (5,1)

Ⓒ (2,5)

2 **Mark the number that has a 6 in the hundreds place.**

Ⓕ 526 Ⓗ 561

Ⓖ 615 Ⓙ 416

4

$$\begin{array}{r} 33 \\ -\ 22 \\ \hline \end{array}$$

Ⓕ 55

Ⓖ 11

Ⓗ 21

Ⓙ 12

Name ______________________________

1 **What shape would come next in the pattern?**

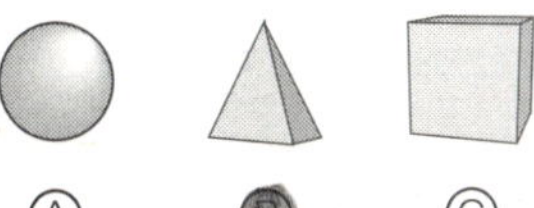

Ⓐ Ⓑ Ⓒ

3

$$\begin{array}{r} 23 \\ -\ 11 \\ \hline \end{array}$$

Ⓐ 11

Ⓑ 12

Ⓒ 34

Ⓓ 9

Ⓔ none of these

2 **Mark the number that tells how many days are in one week.**

Ⓕ 10 Ⓗ 7

Ⓖ 5 Ⓙ 9

4 **1 dime + 2 pennies =**

Ⓕ 12¢ Ⓗ 7¢

Ⓖ 20¢ Ⓙ 14¢

Name ______________________________

$0 + 9 =$

Ⓐ 0

Ⓑ 10

Ⓒ 9

Ⓓ 8

$$\begin{array}{r} 463 \\ -\ 246 \\ \hline \end{array}$$

Ⓕ 217

Ⓖ 223

Ⓗ 227

Ⓙ 709

$$\begin{array}{r} 513 \\ +\ 378 \\ \hline \end{array}$$

Ⓐ 135

Ⓑ 891

Ⓒ 165

Ⓓ 881

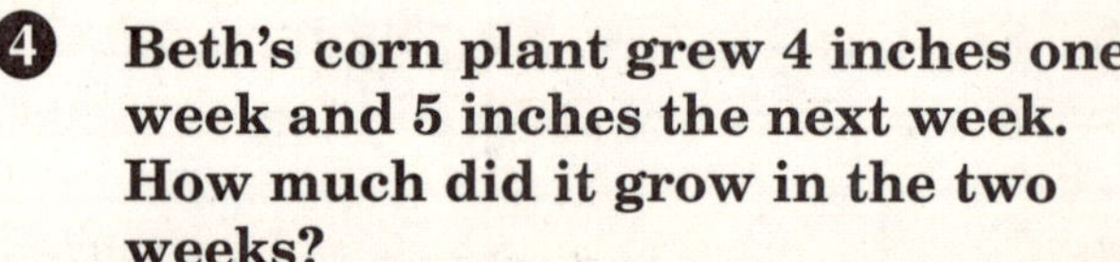
Beth's corn plant grew 4 inches one week and 5 inches the next week. How much did it grow in the two weeks?

Ⓕ 10 inches

Ⓖ 1 inch

Ⓗ 9 inches

Ⓙ 11 inches

Name ________________________________

Mrs. Smith's class was asked to choose a favorite pet. Each student picked only one favorite pet. The graph below shows their choices.

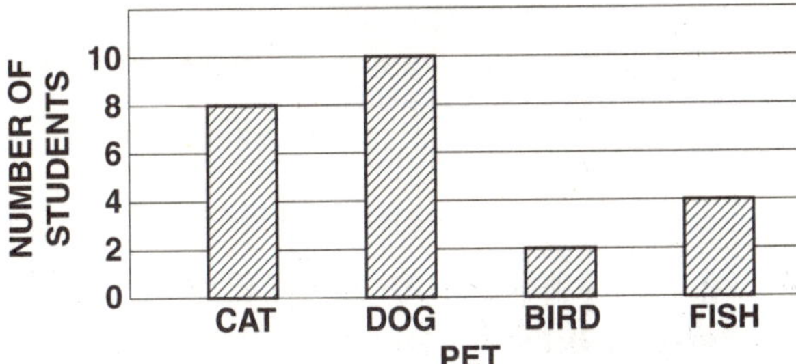

❶ **How many students chose a bird?**

Ⓐ 1 student

Ⓑ 2 students

Ⓒ 4 students

Ⓓ 5 students

❷ **The pet that most students chose was a:**

Ⓕ cat

Ⓖ bird

Ⓗ dog

Ⓙ fish

❸ **Which one of the following is a true statement about the graph on the left?**

Ⓐ More students chose a bird than a fish.

Ⓑ The number of students that chose either a bird or a fish equals the number of students that chose a cat.

Ⓒ Half as many students chose a fish as chose a cat.

Ⓓ The total number of students in the class was 26.

Name ______________________________

1

$$\begin{array}{r} 168 \\ +118 \\ \hline \end{array}$$

Ⓐ 276

Ⓑ 296

Ⓒ 288

Ⓓ 286

2

6 − 5 =

Ⓕ 11

Ⓖ 1

Ⓗ 2

Ⓙ 0

3 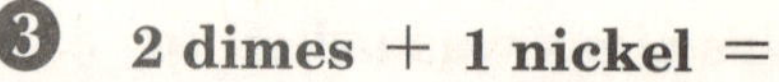

2 dimes + 1 nickel =

Ⓐ 15¢

Ⓑ 21¢

Ⓒ 30¢

Ⓓ 25¢

4 **Mark the fraction that names the shaded part.**

Ⓕ $\frac{1}{4}$

Ⓖ $\frac{1}{3}$

Ⓗ $\frac{1}{2}$

Ⓙ $\frac{2}{3}$

Name ______________________________

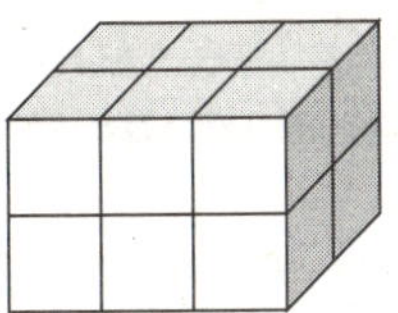

How many small blocks are in this solid shape?

Ⓐ 6 blocks Ⓒ 9 blocks

Ⓑ 12 blocks Ⓓ 14 blocks

$$3 \times 1 =$$

Ⓕ 13 Ⓗ 6

Ⓖ 3 Ⓙ 4

❸ **Which symbol belongs in the ◯ below?**

77  **67**

Ⓐ > Ⓑ < Ⓒ =

$$\begin{array}{r} 52 \\ +\ 17 \\ \hline \end{array}$$

Ⓕ 49

Ⓖ 68

Ⓗ 79

Ⓙ 69

Name ____________________

1 **Which figure below is a triangle?**

Ⓐ 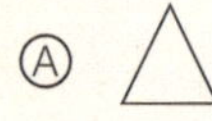

Ⓑ 

Ⓒ 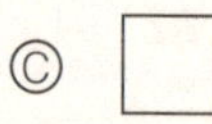

Ⓓ 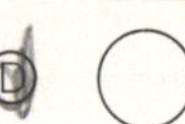

2 **Mary had 9 books. She gave 3 to Jenny. How many does Mary have now?**

Ⓕ 12 books

Ⓖ 5 books

Ⓗ 3 books

Ⓙ 6 books

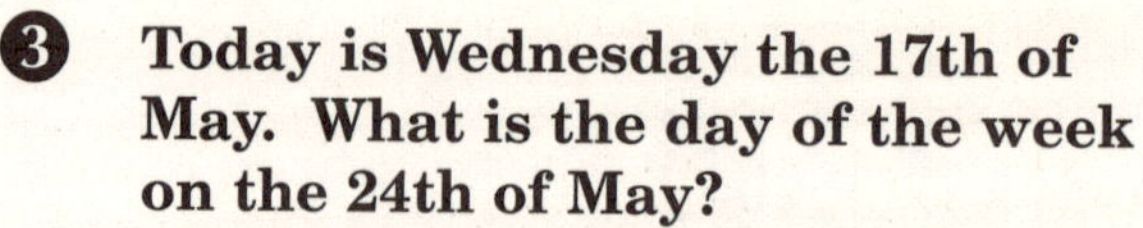

3 **Today is Wednesday the 17th of May. What is the day of the week on the 24th of May?**

Ⓐ Mon.

Ⓑ Tues.

Ⓒ Wed.

Ⓓ Thurs.

4

12 − 8 =

Ⓕ 6

Ⓖ 4

Ⓗ 3

Ⓙ 9

Name ______________________________

1

$3 \times 2 =$

Ⓐ 5
Ⓑ 23
Ⓒ 6
Ⓓ 9

2

1 dime + 1 nickel + 1 penny =

Ⓕ 12¢
Ⓖ 13¢
Ⓗ 16¢
Ⓙ 21¢

3

$$\begin{array}{r} 549 \\ 140 \\ +\ 209 \\ \hline \end{array}$$

Ⓐ 899
Ⓑ 888
Ⓒ 988
Ⓓ 898
Ⓔ none of these

4

$$\begin{array}{r} 819 \\ -\ 298 \\ \hline \end{array}$$

Ⓕ 641
Ⓖ 567
Ⓗ 521
Ⓙ 1,157
Ⓚ none of these

Name ____________________

1 **Mark the number that has a 5 in the tens place.**

Ⓐ 452
Ⓑ 561
Ⓒ 895
Ⓓ 500

2

How long is the piece of ribbon above?

Ⓕ 2 inches
Ⓖ $2\frac{1}{4}$ inches
Ⓗ $2\frac{1}{2}$ inches
Ⓙ $2\frac{3}{4}$ inches

3

$$\begin{array}{r} 60 \\ -\ 10 \\ \hline \end{array}$$

Ⓐ 70
Ⓑ 60
Ⓒ 50
Ⓓ 40

4 **Mark the fraction that names the shaded part.**

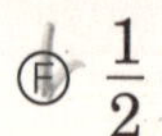

Ⓕ $\frac{1}{2}$

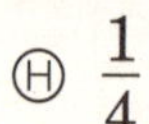

Ⓗ $\frac{1}{4}$

Ⓖ $\frac{1}{3}$

Ⓙ $\frac{3}{4}$

Name ______________________________

$$\begin{array}{r} 4 \\ -\ 0 \\ \hline \end{array}$$

Ⓐ 3

Ⓑ 4

Ⓒ 0

Ⓓ 1

2 **Which symbol belongs in the ◯ below?**

 **48**  **53**

Ⓕ > Ⓖ < Ⓗ =

3 **2 × 4 =**

Ⓐ 24 Ⓒ 4

Ⓑ 8 Ⓓ 7

4 **Grandfather won 6 games of tetherball. Sally won 3. How many <u>more</u> games did Grandfather win?**

Ⓕ 9 games Ⓗ 6 games

Ⓖ 3 games Ⓙ 2 games

Name ____________________

1

$$\begin{array}{r} 460 \\ -\ 325 \\ \hline \end{array}$$

Ⓐ 785

Ⓑ 135

Ⓒ 140

Ⓓ 145

2 **Mark the number that has a 2 in the tens place.**

Ⓕ 421

Ⓖ 142

Ⓗ 241

Ⓙ 442

3

9 + 8 =

Ⓐ 18

Ⓑ 15

Ⓒ 17

Ⓓ 16

4

$$\begin{array}{r} 18 \\ 73 \\ +\ 16 \\ \hline \end{array}$$

Ⓕ 106

Ⓖ 107

Ⓗ 117

Ⓙ 97

Name ______________________________

1 **Tim earns two dollars for every dozen roses he sells. How many dozen roses must he sell to buy a twenty dollar bicycle helmet? Use the table to help you find the answer.**

Dozen roses sold	1	2	3	4	5						
Money earned ($)	2	4	6	8							

Ⓐ 9 dozen roses
Ⓑ 10 dozen roses
Ⓒ 11 dozen roses
Ⓓ 1 dozen roses

2

$$\begin{array}{r} 43 \\ +\ 25 \\ \hline \end{array}$$

Ⓕ 28
Ⓖ 68
Ⓗ 22
Ⓙ 78
Ⓚ none of these

3 **Which figure below is a rectangle?**

Ⓐ (triangle)
Ⓑ (rectangle)
Ⓒ 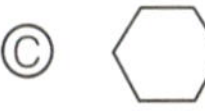
Ⓓ (circle)

Name ______________________________

Look at these numbers: 8, 4, 7, 3

❶ **Write the numbers in order from greatest to least.**

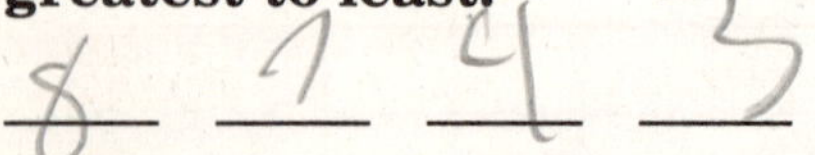

❷ **Using each number one time only, fill in the boxes of this addition problem to make the greatest possible sum.**

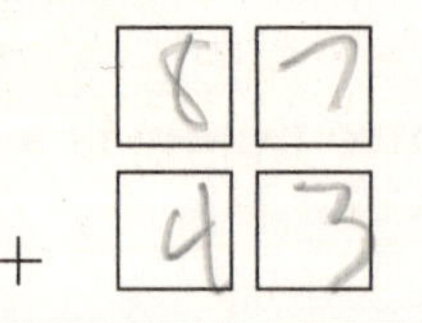

❸ **What is the greatest possible sum?**

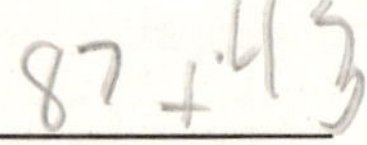

❹ **Using each number one time only, fill in the boxes of this addition problem to make the least possible sum.**

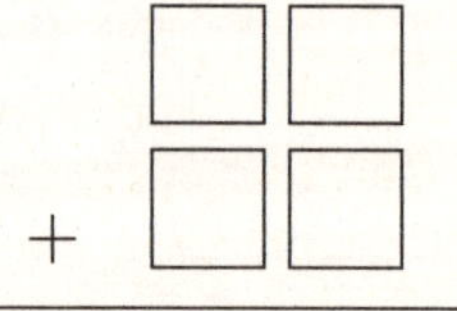

❺ **What is the least possible sum?**

____

Name ____________________

$3 \times 3 =$

Ⓐ 9
Ⓑ 8
Ⓒ 33
Ⓓ 6

2 **Tony ran 2 miles on Monday, 3 miles on Tuesday, and 4 miles on Wednesday. How many miles did he run in all?**

Ⓕ 9 miles
Ⓖ 6 miles
Ⓗ 7 miles
Ⓙ 10 miles

$$\begin{array}{r} 54 \\ +\ 28 \\ \hline \end{array}$$

Ⓐ 84
Ⓑ 72
Ⓒ 82
Ⓓ 74

4 **Mark the fraction that names the shaded part.**

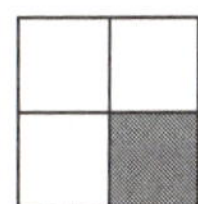

Ⓕ $\frac{1}{2}$
Ⓖ $\frac{1}{3}$
Ⓗ $\frac{1}{4}$
Ⓙ $\frac{1}{8}$

Name ______________________________

5 − 5 =

Ⓐ 10   Ⓒ 1

Ⓑ 0   Ⓓ 5

---

❷

$$\begin{array}{r} 486 \\ -\ 278 \\ \hline \end{array}$$

Ⓕ 208

Ⓖ 212

Ⓗ 215

Ⓙ 218

---

❸ **Guido's dog ate part of his calendar as shown below. Some of the information is missing. Guido is going to Disneyland on the 3rd Saturday of the month. What is the date that he is going to Disneyland?**

S	M	T	W	TH
		1	2	3
6	7	8	9	10
13	14	15	16	17
20	21	22	23	24
27	28	29	30	

Ⓐ 12   Ⓒ 19

Ⓑ 13   Ⓓ 20

Name ______________________________

1

$$\begin{array}{r} 66 \\ +\ 16 \\ \hline \end{array}$$

Ⓐ 76

Ⓑ 82

Ⓒ 80

Ⓓ 72

2 **The temperature at which water boils is shown on the thermometer. What temperature is this?**

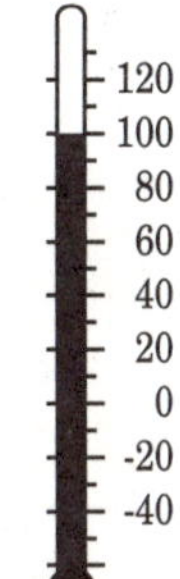

Ⓕ 80°C

Ⓗ 100°C

Ⓖ 90°C

Ⓙ 110°C

3 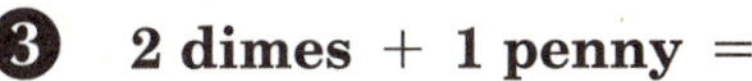

Ⓐ 11¢

Ⓒ 25¢

Ⓑ 3¢

Ⓓ 21¢

4 **Rearrange the digits 3, 5 and 6 to make as many different three-digit numbers as you can.**

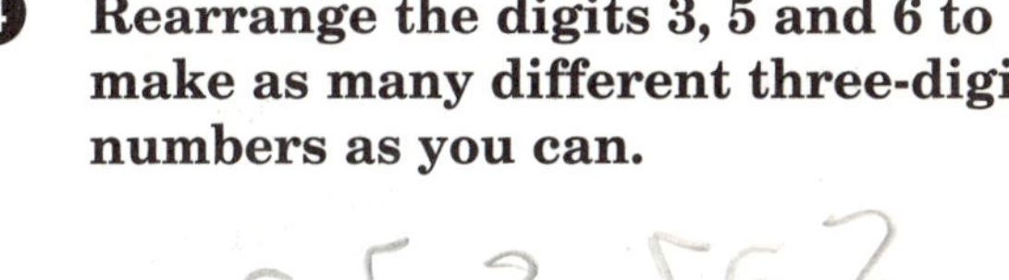

Name ______________________________

1

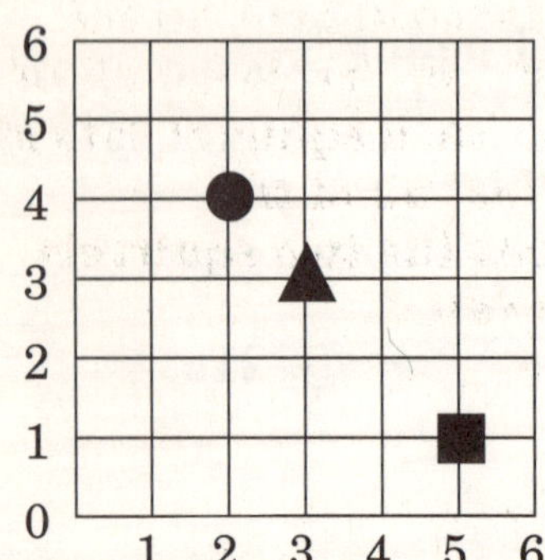

Look at the graph. Where is the ■ ?

Ⓐ (2,4)

Ⓑ (3,3)

Ⓒ (5,1)

2

$$\begin{array}{r} 445 \\ -\ 326 \\ \hline \end{array}$$

Ⓕ 771

Ⓖ 118

Ⓗ 129

Ⓙ 119

Ⓚ none of these

3

Mark the fraction that names the shaded part.

Ⓐ $\frac{1}{2}$

Ⓑ $\frac{1}{3}$

Ⓒ $\frac{1}{4}$

Ⓓ $\frac{2}{3}$

4

Which symbol belongs in the ◯ below?

90  **89**

Ⓕ $>$

Ⓖ $<$ 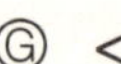

Ⓗ $=$

Name ______________________________

1 **Which figure below is a square?**

Ⓐ 

Ⓑ

Ⓒ

Ⓓ

2

5 + 3 =

Ⓕ 2

Ⓖ 8

Ⓗ 7

Ⓙ 9

3 **There are seven acorns on the ground. Two squirrels eat all of the acorns. Each squirrel eats at least one. List all of the combinations the two squirrels could have eaten.**

Name ____________________

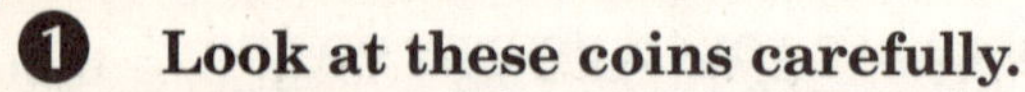

1 **Look at these coins carefully.**

Which set of coins below has the same value as those above?

Ⓐ 

Ⓑ 

Ⓒ 

Ⓓ 

2 **Use an inch ruler and this diagram to help you answer this question. What is the distance between Point A and Point B?**

Point A

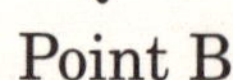

Ⓕ 2 inches

Ⓗ $2\frac{1}{2}$ inches

Ⓖ $2\frac{1}{4}$ inches

Ⓙ $2\frac{3}{4}$ inches

Name ______________________________

1 **Distance from San Francisco, CA to:**

Los Angeles, CA	387 miles
Las Vegas, NV	568 miles
Palm Springs, CA	504 miles
Santa Barbara, CA	336 miles

Which list shows the cities in order from farthest to nearest from San Francisco, CA?

Ⓐ Las Vegas, Palm Springs, Santa Barbara, Los Angeles

Ⓑ Santa Barbara, Los Angeles, Palm Springs, Las Vegas

Ⓒ Las Vegas, Palm Springs, Los Angeles, Santa Barbara

Ⓓ Palm Springs, Las Vegas, Los Angeles, Santa Barbara

2 **Belinda has 26 hair ribbons. She has 8 white, 7 red, 4 black, and 2 green. The rest of the hair ribbons are yellow. How many of the hair ribbons are yellow?**

Ⓕ 4 Ⓗ 6

Ⓖ 5 Ⓙ 3

3 **Today is a warm and sunny day. It would be a good day to go swimming at the lake. What is most likely the temperature today?**

Ⓐ 32° F Ⓒ 85° F

Ⓑ 40° F Ⓓ 132° F

Name ________________________________

1

$3 \times 0 =$

Ⓐ 30 Ⓒ 0

Ⓑ 6 Ⓓ 3

2 **Mark the number that has a 7 in the tens place.**

Ⓕ 701 Ⓗ 874

Ⓖ 427 Ⓙ 742

3

$$\begin{array}{r} 735 \\ 120 \\ +\ 144 \\ \hline \end{array}$$

Ⓐ 984

Ⓑ 991

Ⓒ 999

Ⓓ 878

4 **Sharon did 8 pull-ups on her first try. She did 9 on her next try. How many pull-ups did she do in all?**

Ⓕ 16 pull-ups Ⓗ 19 pull-ups

Ⓖ 18 pull-ups Ⓙ 17 pull-ups

Name ______________________________

1

$$\begin{array}{r} 487 \\ +\ 369 \\ \hline \end{array}$$

Ⓐ 118
Ⓑ 855
Ⓒ 866
Ⓓ 846
Ⓔ none of these

2

$$\begin{array}{r} \$\ \ 6.15 \\ -\ \ 4.03 \\ \hline \end{array}$$

Ⓕ $4.12
Ⓖ $10.02
Ⓗ $2.02
Ⓙ $2.12
Ⓚ none of these

3 **What time is shown?**

Ⓐ 12:15
Ⓑ 12:03
Ⓒ 3:00
Ⓓ 3:12

4 **Mark the number that tells how many months are in one year.**

Ⓕ 10
Ⓖ 12
Ⓗ 15
Ⓙ 7

Name ______________________________

1

```
  234
− 215
```

Ⓐ 449

Ⓑ 29

Ⓒ 19

Ⓓ 18

2

```
  36
+ 15
```

Ⓕ 41

Ⓖ 21

Ⓗ 51

Ⓙ 24

3 **Mr. Perez asked his 4th grade class to vote for one color to paint his chair. Every student voted for only one color. The graph below shows the students' choices.**

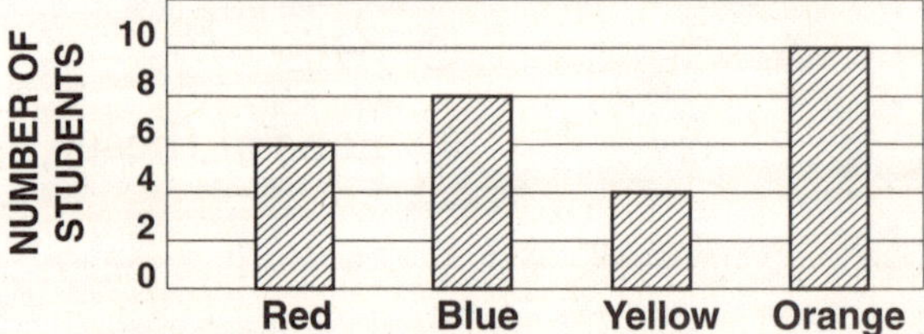

Find the statement that is not true about the students' choices above.

Ⓐ Twice as many students liked blue as liked yellow.

Ⓑ There are 26 students in the class.

Ⓒ More students liked red than yellow.

Ⓓ 2 more students liked orange than blue.

Name ______________________________

1 **Which symbol belongs in the ◯ below?**

17 ◯ **17 − 0**

Ⓐ > Ⓑ < Ⓒ =

2 **596 =**

Ⓕ _6_ hundreds _9_ tens _5_ ones

Ⓖ _5_ hundreds _6_ tens _9_ ones

Ⓗ _5_ hundreds _6_ tens _5_ ones

Ⓙ _5_ hundreds _9_ tens _6_ ones

3

$$\begin{array}{r} 420 \\ 365 \\ +\ 113 \\ \hline \end{array}$$

Ⓐ 988

Ⓑ 987

Ⓒ 898

Ⓓ 908

4 **Mark the even number.**

Ⓕ 3 Ⓗ 6

Ⓖ 5 Ⓙ 1

Name ________________________________

1 **Which is the numeral for 3 hundreds 4 tens 5 ones?**

Ⓐ 543 Ⓒ 345

Ⓑ 354 Ⓓ 305

3 **Which figure below is a circle?**

Ⓐ ○ Ⓒ □

Ⓑ △ Ⓓ ▭

2

$$\begin{array}{r} 317 \\ +\ 201 \\ \hline \end{array}$$

Ⓕ 116

Ⓖ 618

Ⓗ 118

Ⓙ 518

4 **4 dimes + 1 nickel =**

Ⓕ 9¢ Ⓗ 45¢

Ⓖ 41¢ Ⓙ 21¢

Name ______________________________

1

$$\begin{array}{r} 971 \\ -\ 765 \\ \hline \end{array}$$

Ⓐ 206

Ⓑ 1,736

Ⓒ 216

Ⓓ 294

2 **Mark the even number.**

Ⓕ 3

Ⓖ 8

Ⓗ 9

Ⓙ 7

3

$$\begin{array}{r} \$\ 3.41 \\ +\ 4.27 \\ \hline \end{array}$$

Ⓐ $8.68

Ⓑ $7.78

Ⓒ $7.66

Ⓓ $7.68

4 **What time is shown?**

Ⓕ 6:09

Ⓖ 10:30

Ⓗ 9:30

Ⓙ 6:48

Name ______________________________

1 **There were 27 people on the bus. Five people got off at Main Street. How many people remained on the bus?**

Ⓐ 32 people
Ⓑ 37 people
Ⓒ 22 people
Ⓓ 21 people

2 **Which is the numeral for 6 hundreds 8 tens 4 ones?**

Ⓕ 468
Ⓖ 864
Ⓗ 6084
Ⓙ 684

3

$$2 \times 5 =$$

Ⓐ 25
Ⓑ 10
Ⓒ 12
Ⓓ 7

4

$$\begin{array}{r} 505 \\ 362 \\ +\ 131 \\ \hline \end{array}$$

Ⓕ 1008
Ⓖ 998
Ⓗ 997
Ⓙ 908
Ⓚ none of these

Name ______________________________

$$\begin{array}{r} 599 \\ -\ 306 \\ \hline \end{array}$$

Ⓐ 203

Ⓑ 202

Ⓒ 293

Ⓓ 192

2 **Mark the number that has a 6 in the hundreds place.**

Ⓕ 1726 Ⓗ 1634

Ⓖ 1462 Ⓙ 6274

3

$$\begin{array}{r} 24 \\ +\ 28 \\ \hline \end{array}$$

Ⓐ 42

Ⓑ 52

Ⓒ 56

Ⓓ 44

4

$4 \times 2 =$

Ⓕ 6 Ⓗ 8

Ⓖ 10 Ⓙ 12

Name ______________________________

1 **Mark the fraction that names the shaded part.**

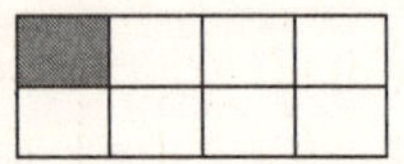

Ⓐ $\frac{1}{4}$ Ⓑ $\frac{1}{3}$ Ⓒ $\frac{1}{8}$ Ⓓ $\frac{1}{6}$

2 **Which shows a diagonal line?**

Ⓕ 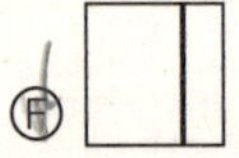

Ⓗ 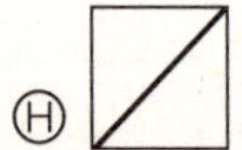

Ⓖ 

Ⓙ 

3

$$\begin{array}{r} 313 \\ 475 \\ +\ 109 \\ \hline \end{array}$$

Ⓐ 898

Ⓑ 888

Ⓒ 887

Ⓓ 897

4 **845 =**

Ⓕ _5_ hundreds _4_ tens _8_ ones

Ⓖ _4_ hundreds _8_ tens _5_ ones

Ⓗ _8_ hundreds _4_ tens _5_ ones

Ⓙ _8_ hundreds _40_ tens _5_ ones

Name ________________________________

$$\begin{array}{r} \$\ 5.16 \\ +\ 3.72 \\ \hline \end{array}$$

Ⓐ $8.68

Ⓑ $8.88

Ⓒ $2.64

Ⓓ $8.98

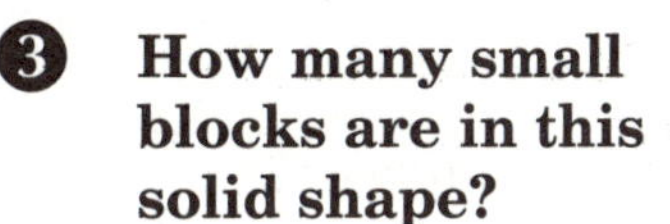

3 **How many small blocks are in this solid shape?**

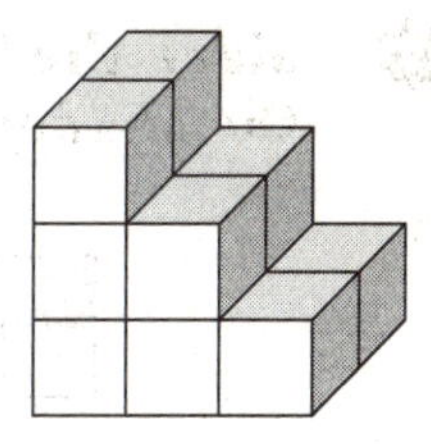

Ⓐ 10 blocks

Ⓒ 12 blocks

Ⓑ 14 blocks

Ⓓ 16 blocks

2

$$\begin{array}{r} 900 \\ -\ 360 \\ \hline \end{array}$$

Ⓕ 660

Ⓖ 560

Ⓗ 540

Ⓙ 1,260

4 **Mark the odd number.**

Ⓕ 2

Ⓗ 10

Ⓖ 9

Ⓙ 8

Name ______________________________

1

$$\begin{array}{r} 367 \\ 394 \\ +\ 226 \\ \hline \end{array}$$

Ⓐ 977

Ⓑ 887

Ⓒ 886

Ⓓ 987

2 **Emily is 10 years old. Her big sister is 23 years old. How many years younger is Emily?**

Ⓕ 33 years　　Ⓗ 3 years

Ⓖ 20 years　　Ⓙ 13 years

3

$$2 \times \square = 2$$

Ⓐ 0　　Ⓒ 2

Ⓑ 1　　Ⓓ 4

4 **Which figure below is a triangle?**

Ⓕ □　　Ⓗ ▭

Ⓖ 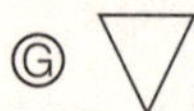　　Ⓙ ○

Name ______________________________

1

$$\begin{array}{r} 31 \\ -\ 19 \\ \hline \end{array}$$

Ⓐ 28

Ⓑ 18

Ⓒ 12

Ⓓ 22

Ⓔ none of these

2

$$\begin{array}{r} 3 \\ \times\ 4 \\ \hline \end{array}$$

Ⓕ 7

Ⓖ 15

Ⓗ 20

Ⓙ 12

Ⓚ none of these

3 **Tommy the turtle was 30 years old in 1991. In which year did Tommy hatch?**

Ⓐ 1988

Ⓑ 1691

Ⓒ 1961

Ⓓ 2021

4 **Mark the odd number.**

Ⓕ 12

Ⓖ 13

Ⓗ 14

Ⓙ 16

Name ______________________

1 **What time is shown?**

Ⓐ 12:35

Ⓑ 12:07

Ⓒ 7:12

Ⓓ 7:00

2 **Which is the numeral for 9 hundreds 9 tens 3 ones?**

Ⓕ 9903

Ⓖ 399

Ⓗ 9930

Ⓙ 993

3

$$\begin{array}{r} \mathbf{49} \\ + \ \mathbf{21} \\ \hline \end{array}$$

Ⓐ 70

Ⓑ 28

Ⓒ 60

Ⓓ 61

4

$$\begin{array}{r} \mathbf{758} \\ - \ \mathbf{306} \\ \hline \end{array}$$

Ⓕ 1064

Ⓖ 452

Ⓗ 402

Ⓙ 358

Name ______________________________

1

$$\begin{array}{r} 42 \\ -\ 16 \\ \hline \end{array}$$

Ⓐ 34

Ⓑ 36

Ⓒ 26

Ⓓ 54

2 **What is the perimeter of the figure below?**

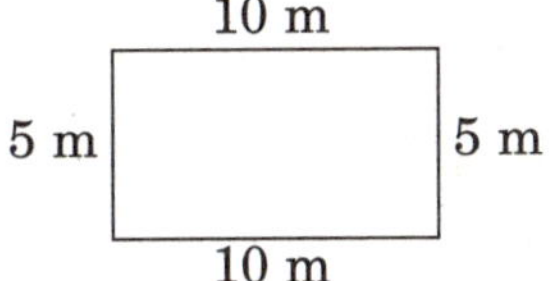

Ⓕ 15 m

Ⓖ 25 m

Ⓗ 30 m

Ⓙ 35 m

3 **There are eight worms on the grass. Three birds eat all of the worms. Each bird eats at least one. List all of the combinations the three birds could have eaten.**

Name ____________________

1 **At Lake School there are 435 children. There are 223 boys. How many girls are there?**

Ⓐ 658 girls

Ⓑ 202 girls

Ⓒ 112 girls

Ⓓ 212 girls

2 **Which figure is a parallelogram?**

Ⓕ 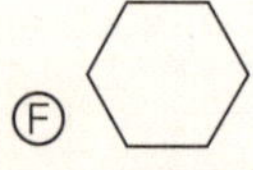

Ⓗ 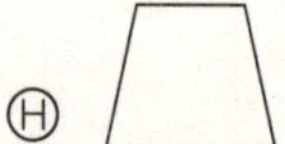

Ⓖ 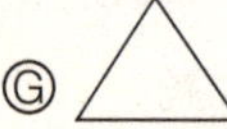

Ⓙ 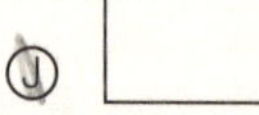

3

$$\begin{array}{r} 647 \\ -\ 520 \\ \hline \end{array}$$

Ⓐ 127

Ⓑ 120

Ⓒ 1167

Ⓓ 227

4

$$\begin{array}{r} 0 \\ \times\ 5 \\ \hline \end{array}$$

Ⓕ 5

Ⓖ 50

Ⓗ 0

Ⓙ 10

Name ______________________________________

The drawing above is a:

Ⓐ right angle
Ⓑ ray
Ⓒ line
Ⓓ line segment

$$\begin{array}{r} \$\ 4.77 \\ -\ 2.36 \\ \hline \end{array}$$

Ⓕ $2.41
Ⓖ $6.41
Ⓗ $6.30
Ⓙ $2.31

3 **Which symbol belongs in the ○ below?**

15  **8 + 8**

Ⓐ > Ⓑ < Ⓒ =

4 **Mark the number that has a 3 in the ones place.**

Ⓕ 1341
Ⓖ 2834
Ⓗ 6403
Ⓙ 3604

Name ______________________________

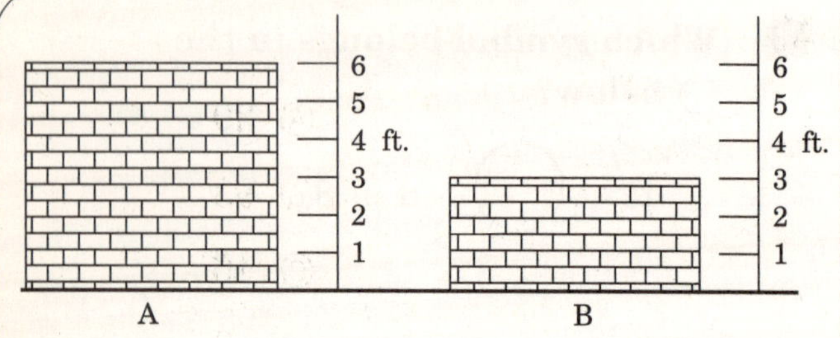

❶ **How much taller is wall A than wall B?**

Ⓐ 2 feet
Ⓑ $2\frac{1}{2}$ feet
Ⓒ 3 feet
Ⓓ $3\frac{1}{2}$ feet

❷ **If you put wall B on top of wall A, how tall would the new wall be?**

Ⓕ 7 feet
Ⓖ 8 feet
Ⓗ 9 feet
Ⓙ 10 feet

❸ **Mark the fraction that names the shaded part.**

Ⓐ $\frac{1}{2}$ Ⓑ $\frac{2}{4}$ Ⓒ $\frac{2}{3}$ Ⓓ $\frac{1}{3}$

❹

$$\begin{array}{r} 68 \\ +\ 18 \\ \hline \end{array}$$

Ⓕ 76
Ⓖ 86
Ⓗ 50
Ⓙ 78
Ⓚ none of these

Name ______________________________

1

$$\begin{array}{r} 755 \\ +\ 169 \\ \hline \end{array}$$

Ⓐ 614

Ⓑ 925

Ⓒ 914

Ⓓ 924

2 **Mother baked 24 muffins. The family ate 10 of them for dinner. How many are left?**

Ⓕ 34 muffins

Ⓗ 14 muffins

Ⓖ 10 muffins

Ⓙ 24 muffins

3

$$\begin{array}{r} 13 \\ \times\ \ 3 \\ \hline \end{array}$$

Ⓐ 39

Ⓑ 93

Ⓒ 16

Ⓓ 19

4 **Mark the odd number.**

Ⓕ 7

Ⓗ 8

Ⓖ 14

Ⓙ 10

Name ______________________________

Look at these numbers: 1, 5, 6, 2

❶ **Write the numbers in order from greatest to least.**

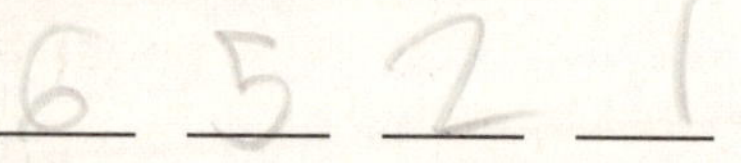

❷ **Using each number one time only, fill in the boxes of this addition problem to make the greatest possible sum.**

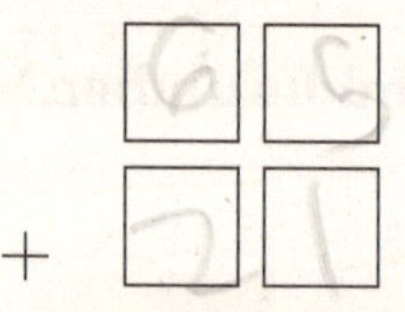

❸ **What is the greatest possible sum?**

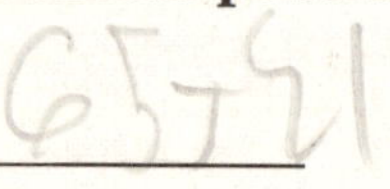

❹ **Using each number one time only, fill in the boxes of this addition problem to make the least possible sum.**

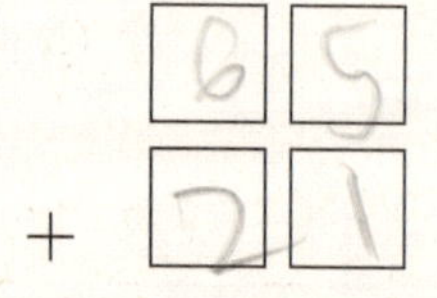

❺ **What is the least possible sum?**

Name ______________________________

1

$$\begin{array}{r} 430 \\ -\ \ 80 \\ \hline \end{array}$$

Ⓐ 510

Ⓑ 450

Ⓒ 340

Ⓓ 350

2 **Mark the number that tells how many months are in one year.**

Ⓕ 10

Ⓖ 12

Ⓗ 7

Ⓙ 52

3

$$\begin{array}{r} \$\ \ 8.23 \\ +\ \ 1.66 \\ \hline \end{array}$$

Ⓐ $9.49

Ⓑ $10.89

Ⓒ $7.43

Ⓓ $9.89

4 **What time is shown?**

Ⓕ 11:30

Ⓖ 6:00

Ⓗ 12:30

Ⓙ 6:11

Name ______________________________

1 **Look at these coins carefully.**

Which set of coins below has the same value as those above?

Ⓐ 

Ⓑ 

Ⓒ 

Ⓓ 

2 **Dave and Monica ate cookies after dinner. Dave ate 3 cookies. What else do you need to know to find the number of cookies they ate altogether?**

Ⓕ What kind of cookies are they?

Ⓖ Who made the cookies?

Ⓗ How many cookies were made?

Ⓙ How many cookies Monica ate?

3 **Scott has 45 toy race cars. He has 12 red, 10 blue, 8 green, 6 white, and 4 black. The rest of the race cars are purple. How many of the race cars are purple?**

Ⓐ 3

Ⓑ 4

Ⓒ 5

Ⓓ 6

Name ____________________

1 **Which of the shaded figures has the greatest shaded area?**

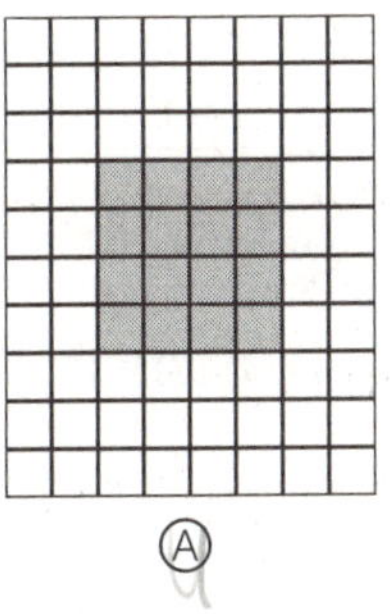

Ⓐ

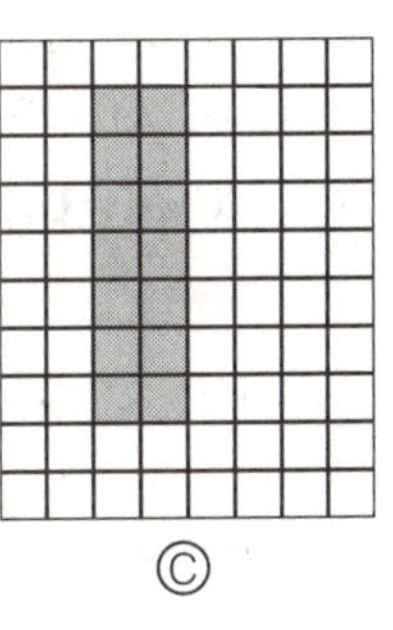

Ⓒ

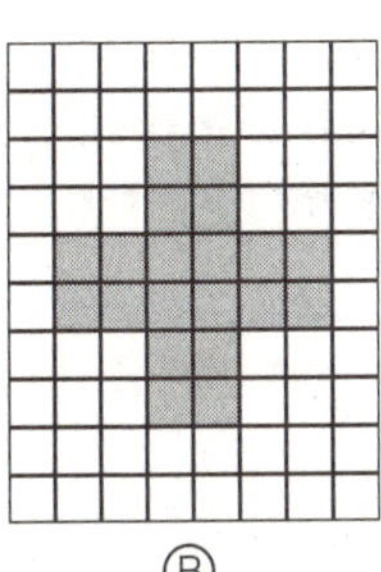

Ⓑ

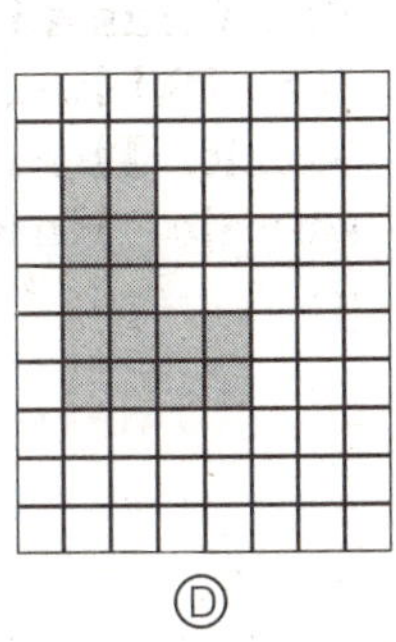

Ⓓ

2 **Use a centimeter ruler and this diagram to help you answer the question. What is the distance between Point A and Point B?**

Point A

Point B

Ⓕ 4 cm

Ⓖ 5 cm

Ⓗ 6 cm

Ⓙ 7 cm

Name ______________________________

$$\begin{array}{r} 23 \\ \times \ \ 2 \\ \hline \end{array}$$

Ⓐ 46

Ⓑ 64

Ⓒ 44

Ⓓ 25

Ⓔ none of these

2 **What shape would come next in the pattern?**

Ⓕ  Ⓖ  Ⓗ 

$$\begin{array}{r} 83 \\ -\ 38 \\ \hline \end{array}$$

Ⓐ 55

Ⓑ 45

Ⓒ 121

Ⓓ 35

Ⓔ none of these

4 **Ann got a book from the library. She may keep it for 2 weeks. How many days may she keep it?**

Ⓕ 10 days

Ⓖ 12 days

Ⓗ 20 days

Ⓙ 14 days

Name ________________________________

1

$$\begin{array}{r} 23 \\ \times \quad 3 \\ \hline \end{array}$$

Ⓐ 66

Ⓑ 69

Ⓒ 26

Ⓓ 606

2

$$\begin{array}{r} \$\,17.35 \\ +\,32.64 \\ \hline \end{array}$$

Ⓕ $48.09

Ⓖ $50.89

Ⓗ $49.99

Ⓙ $45.31

3

708 =

Ⓐ __7__ hundreds __10__ tens __8__ ones

Ⓑ __7__ hundreds __0__ tens __8__ ones

Ⓒ __8__ hundreds __0__ tens __7__ ones

Ⓓ __7__ hundreds __8__ tens __0__ ones

4 **Today is Monday. Bob's birthday is tomorrow. Mark the day of Bob's birthday.**

Ⓕ Tuesday

Ⓖ Wednesday

Ⓗ Sunday

Ⓙ Saturday

Name ______________________________

1

$$\begin{array}{r} 31 \\ \times \quad 3 \\ \hline \end{array}$$

Ⓐ 34

Ⓑ 91

Ⓒ 63

Ⓓ 93

3 **18 rounded to the nearest ten is:**

Ⓐ 0

Ⓑ 10

Ⓒ 20

Ⓓ 15

2

$$\begin{array}{r} 346 \\ -\ 227 \\ \hline \end{array}$$

Ⓕ 573

Ⓖ 119

Ⓗ 121

Ⓙ 19

4

16 − 8 =

Ⓕ 8

Ⓖ 12

Ⓗ 4

Ⓙ 6

Name ______________________________

1

$$\begin{array}{r} 733 \\ 120 \\ +\ 144 \\ \hline \end{array}$$

Ⓐ 1007

Ⓑ 987

Ⓒ 997

Ⓓ 887

2 **Mark the number with the greatest value.**

Ⓕ 101

Ⓖ 100

Ⓗ 111

Ⓙ 11

3

$$2\overline{)4}$$

Ⓐ 2

Ⓑ 8

Ⓒ 4

Ⓓ 6

4 **41 rounded to the nearest ten is:**

Ⓕ 40

Ⓖ 45

Ⓗ 30

Ⓙ 50

Name ______________________

1

$3\overline{)6}$

Ⓐ 9
Ⓑ 3
Ⓒ 6
Ⓓ 2

2 **Carlos read 33 pages of his book one day and 25 pages the next. How many pages did he read in the two days?**

Ⓕ 18 pages
Ⓖ 12 pages
Ⓗ 58 pages
Ⓙ 68 pages

3 **What is the missing number?**

10, 15, _____, 25

Ⓐ 20
Ⓑ 21
Ⓒ 30
Ⓓ 35

4

$$\begin{array}{r} 721 \\ -\ 420 \\ \hline \end{array}$$

Ⓕ 1141
Ⓖ 301
Ⓗ 321
Ⓙ 300

Name ______________________________

$$\begin{array}{r} 22 \\ \times \quad 2 \\ \hline \end{array}$$

Ⓐ 24
Ⓑ 44
Ⓒ 42
Ⓓ 202
Ⓔ none of these

$$\begin{array}{r} 456 \\ +\ 235 \\ \hline \end{array}$$

Ⓕ 221
Ⓖ 693
Ⓗ 681
Ⓙ 691
Ⓚ none of these

3 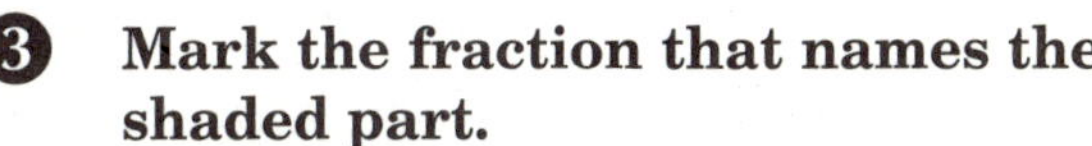

Mark the fraction that names the shaded part.

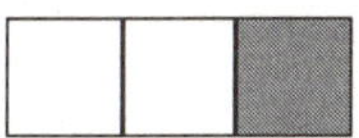

Ⓐ $\frac{1}{3}$ Ⓑ $\frac{1}{2}$ Ⓒ $\frac{2}{3}$ Ⓓ $\frac{1}{4}$

4 

Which is the numeral for six hundred fifty-three?

Ⓕ 6503 Ⓗ 653
Ⓖ 563 Ⓙ 365

Name ______________________

1

$$\begin{array}{r} 547 \\ -\ 318 \\ \hline \end{array}$$

Ⓐ 231

Ⓑ 229

Ⓒ 131

Ⓓ 239

3

2 + 9 = 9 + _____

Ⓐ 11

Ⓑ 9

Ⓒ 2

Ⓓ 12

2 **Estimate the answer by rounding.**

$$\begin{array}{r} 11 \\ +\ 31 \\ \hline \end{array}$$

Ⓕ 20

Ⓖ 40

Ⓗ 30

Ⓙ 50

4

$5\overline{)0}$

Ⓕ 5

Ⓖ 0

Ⓗ 1

Ⓙ 2

Name ______________________________

1

$$\begin{array}{r} \$\,24.23 \\ +\,14.65 \\ \hline \end{array}$$

Ⓐ $10.42

Ⓑ $48.87

Ⓒ $38.88

Ⓓ $38.98

2 **Estimate the answer by rounding.**

$$\begin{array}{r} 19 \\ +\,32 \\ \hline \end{array}$$

Ⓕ 40

Ⓖ 60

Ⓗ 50

Ⓙ 70

3 **Which symbol belongs in the ◯ below?**

20 ◯ **2 × 10**

Ⓐ > Ⓑ < Ⓒ =

4 **There are 6 baseball teams in the tournament. Each team has 9 players. How many players are there in all?**

Ⓕ 54 players Ⓗ 36 players

Ⓖ 45 players Ⓙ 63 players

Name ______________________________

1 **How many days are in 2 weeks?**

Ⓐ 20 days
Ⓑ 10 days
Ⓒ 14 days
Ⓓ 4 days

2

$$\begin{array}{r} \$\ 7.47 \\ -\ 6.47 \\ \hline \end{array}$$

Ⓕ $1.00
Ⓖ $ .10
Ⓗ $13.94
Ⓙ $1.47

3

200 + 90 + 2 =

Ⓐ 200,902
Ⓑ 292
Ⓒ 2092
Ⓓ 2902

4

$$\begin{array}{r} 43 \\ \times\ \ 2 \\ \hline \end{array}$$

Ⓕ 45
Ⓖ 44
Ⓗ 86
Ⓙ 85

Name ______________________________

The graph below shows the number of birds that flew over the playground between 9:00 A.M. and 2:00 P.M.

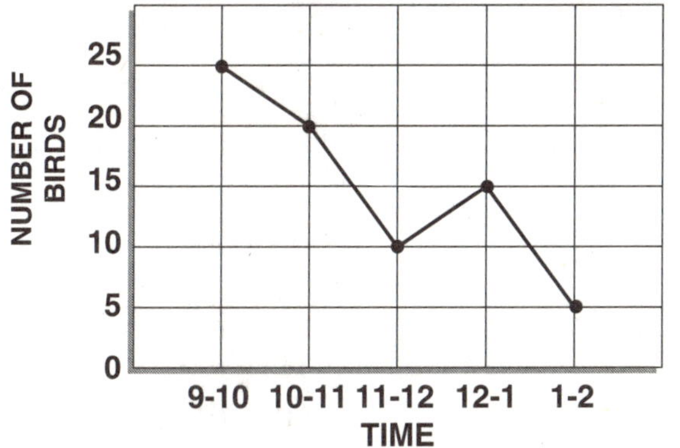

1 How many more birds were counted in the morning than in the afternoon?

Ⓐ 55 birds Ⓒ 35 birds

Ⓑ 20 birds Ⓓ 40 birds

2 In which hour did the fewest birds fly over the playground?

Ⓕ 10 – 11 Ⓗ 12 – 1

Ⓖ 11 – 12 Ⓙ 1 – 2

3 In which hour did the number of birds increase compared to the hour just before it?

Ⓐ 10 – 11 Ⓒ 12 – 1

Ⓑ 11 – 12 Ⓓ 1 – 2

Name ______________________________

1

$$\begin{array}{r} 513 \\ +\ 278 \\ \hline \end{array}$$

Ⓐ 365

Ⓑ 791

Ⓒ 781

Ⓓ 785

Ⓔ none of these

3

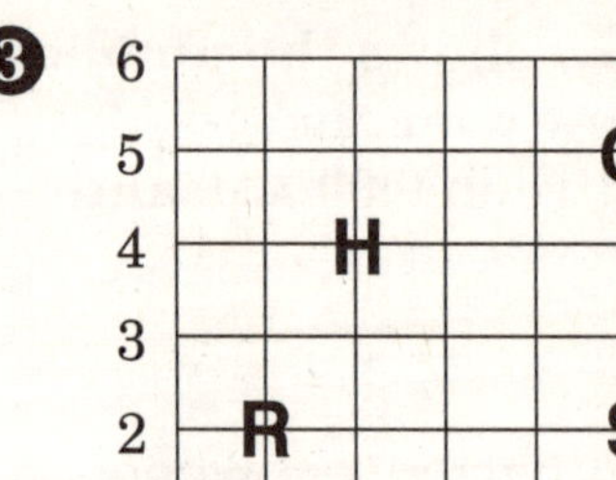

Look at the graph. Where is the H located?

Ⓐ (4, 5)

Ⓑ (5, 2)

Ⓒ (2, 4)

Ⓓ (1, 2)

2 **Alberto has 43 colored pencils. Cindy has 30. How many more than Cindy does Alberto have?**

Ⓕ 73 colored pencils

Ⓖ 10 colored pencils

Ⓗ 3 colored pencils

Ⓙ 13 colored pencils

4

300 + 60 + 3 =

Ⓕ 30,603

Ⓖ 363

Ⓗ 3603

Ⓙ 633

Name ______________________________

1

$$\begin{array}{r} 321 \\ \times \quad 2 \\ \hline \end{array}$$

Ⓐ 323

Ⓑ 641

Ⓒ 751

Ⓓ 642

2 **What time is shown?**

Ⓕ 3:11

Ⓖ 11:03

Ⓗ 3:55

Ⓙ 11:15

3 **Mark the even number.**

Ⓐ 9

Ⓑ 22

Ⓒ 37

Ⓓ 31

4 **Rover found 2 bones. Blake found bones. How many bones in all? What else do you need to know to answer the question?**

Ⓕ What kind of bones they were?

Ⓖ How many bones Blake found?

Ⓗ How many bones were left?

Name ______________________________

1

$$\begin{array}{r} \$\ 6.29 \\ -\ 4.17 \\ \hline \end{array}$$

Ⓐ $10.46

Ⓑ $2.02

Ⓒ $2.19

Ⓓ $2.12

2 **11 rounded to the nearest ten is:**

Ⓕ 1

Ⓗ 15

Ⓖ 10

Ⓙ 20

3 **There are 7 days in a week. How many days are there in 9 weeks?**

Ⓐ 56 days

Ⓒ 82 days

Ⓑ 63 days

Ⓓ 81 days

4 **Today is Friday. Tomorrow we are going to Hawaii. Mark the day we will go to Hawaii.**

Ⓕ Thursday

Ⓗ Sunday

Ⓖ Saturday

Ⓙ Monday

Name ______________________________

$3 + 5 = \_\_\_\_\_ + 3$

Ⓐ 5 Ⓒ 2

Ⓑ 8 Ⓓ 0

2 **What is the perimeter of the figure below?**

6 m

6 m

Ⓕ 12 m Ⓗ 32 m

Ⓖ 24 m Ⓙ 36 m

3 **In a baseball game, 9 fly balls were hit into the outfield. The three outfielders caught all 9. No outfielder caught more than 5. List all of the combinations the three outfielders could have caught.**

I do not

no

Name ________________________________

1 **Shade a part of the figure to show what $\frac{1}{4}$ means.**

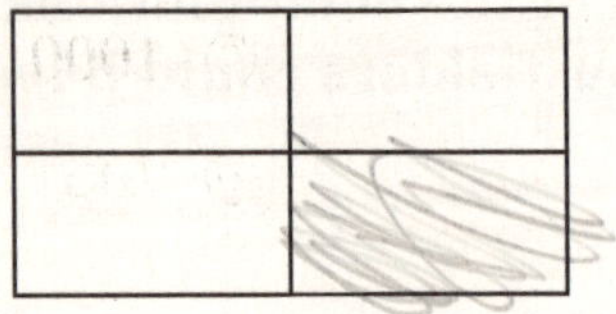

2 **Shade a part of the figure to show what $\frac{1}{2}$ means.**

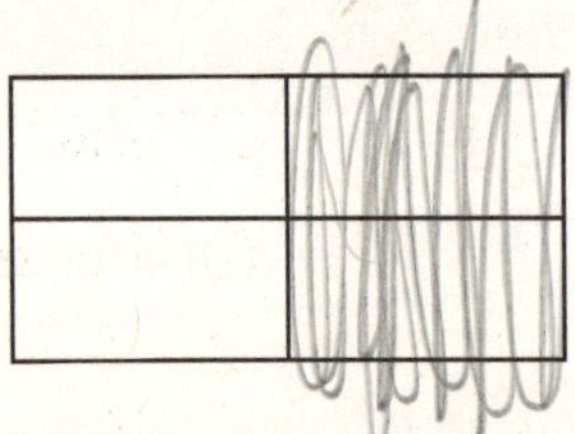

3 **Shade a part of the figure to show what $\frac{1}{4} + \frac{1}{2}$ means.**

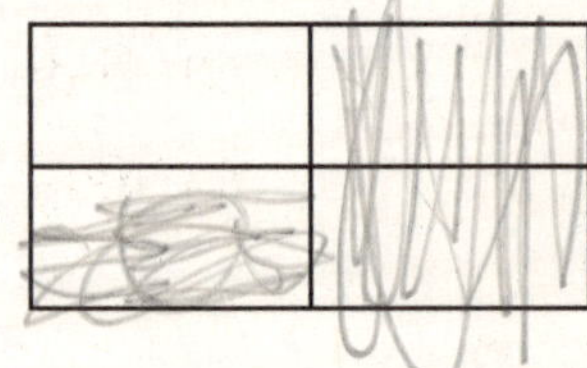

4 **Use your picture from problem 3 to write the fraction that is equal to $\frac{1}{4} + \frac{1}{2}$.**

Name ______________________________

1 **Which is the numeral for four hundred thirty?**

Ⓐ 4030 Ⓒ 403

Ⓑ 4300 Ⓓ 430

2

327
+ 256

Ⓕ 573
Ⓖ 574
Ⓗ 582
Ⓙ 683
Ⓚ none of these

3 **890 rounded to the nearest hundred is:**

Ⓐ 900 Ⓒ 1000

Ⓑ 800 Ⓓ 700

4

The drawing above is a:

Ⓕ right angle Ⓗ line

Ⓖ ray Ⓙ line segment

Name ____________________

1 **Mark the odd number.**

Ⓐ 34 Ⓒ 36

Ⓑ 35 Ⓓ 30

2 **What time is shown?**

Ⓕ 1:15 Ⓗ 3:07

Ⓖ 1:03 Ⓙ 3:01

3

$$\begin{array}{r} 226 \\ -\ 108 \\ \hline \end{array}$$

Ⓐ 18

Ⓑ 122

Ⓒ 118

Ⓓ 128

4 **Which symbol belongs in the ○ below?**

$$\frac{1}{2} \quad \bigcirc \quad \frac{1}{3}$$

Ⓕ > Ⓖ < Ⓗ =

Name ________________________________

1 **The pictograph below shows the number of tennis rackets a store sold during a 4 day period.**

DAYS OF THE WEEK	NUMBER SOLD
Thursday	🎾 🎾 🎾
Friday	🎾 🎾 🎾 🎾 🎾
Saturday	🎾 🎾 🎾 🎾 🎾 🎾 🎾 🎾
Sunday	🎾 🎾 🎾 🎾 🎾 🎾
🎾 = 10 Tennis Rackets	

How many tennis rackets were sold on Sunday?

Ⓐ 8 tennis rackets

Ⓑ 22 tennis rackets

Ⓒ 6 tennis rackets

Ⓓ 60 tennis rackets

2

$$\begin{array}{r} \$\ 8.37 \\ -\ 3.16 \\ \hline \end{array}$$

Ⓕ $5.11

Ⓖ $4.21

Ⓗ $5.21

Ⓙ $5.19

3 **What is the missing number?**

4, ___, 12, 16

Ⓐ 5

Ⓒ 6

Ⓑ 9

Ⓓ 8

Name ______________________________

❶ $3 + (6 + 5) = (3 + 6) +$ _____

Ⓐ 9

Ⓑ 14

Ⓒ 5

Ⓓ 11

❷ **Ryan has 6 coins which total 81¢. Which of the following is the correct combination of coins to give Ryan exactly 81¢?**

Ⓕ 3 quarters and 3 pennies

Ⓖ 2 quarters, 3 dimes, and 1 penny

Ⓗ 2 quarters, 1 dime, and 1 penny

Ⓙ 2 quarters, 2 dimes, and 2 nickels

❸

$6\overline{)6}$

Ⓐ 0

Ⓑ 1

Ⓒ 6

Ⓓ 36

❹

$$\begin{array}{r} 232 \\ \times \quad 3 \\ \hline \end{array}$$

Ⓕ 235

Ⓖ 496

Ⓗ 596

Ⓙ 696

Name ______________________________

1 **Which of the following figures is a cube?**

Ⓐ 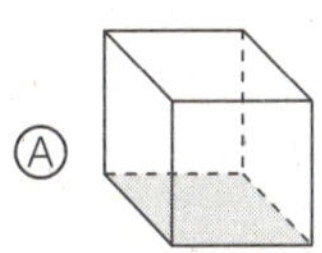

Ⓒ 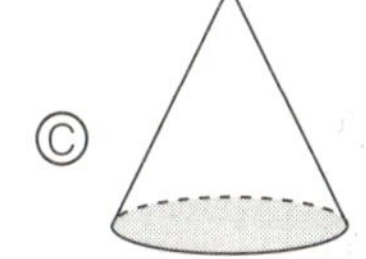

Ⓑ 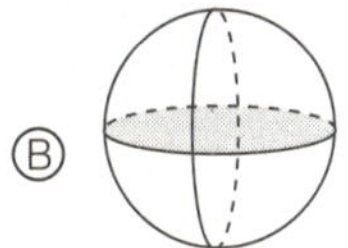

Ⓓ 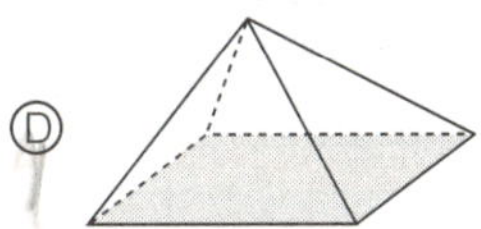

2

$$
\begin{array}{r}
\$\ 3.46 \\
+\ 4.18 \\
\hline
\end{array}
$$

Ⓕ $8.54

Ⓖ $8.64

Ⓗ $7.54

Ⓙ $7.64

Ⓚ none of these

3 **675 rounded to the nearest hundred is:**

Ⓐ 600

Ⓒ 700

Ⓑ 650

Ⓓ 800

4 **Which is the number with the least value?**

Ⓕ 53

Ⓗ 33

Ⓖ 35

Ⓙ 38

Name ______________________________

❶ **At the annual company picnic, there were 308 full-time employees, 398 part-time employees, and 403 guests. *About* how many people were at the picnic?**

Ⓐ 1100 people
Ⓑ 1000 people
Ⓒ 900 people
Ⓓ 950 people

❷ **There are 66 posts on a fence. Tom has painted 12 posts per day for the last 3 days. How many posts does he have left to paint?**

Ⓕ 54 posts
Ⓖ 36 posts
Ⓗ 30 posts
Ⓙ 20 posts

❸ **Which of the shaded figures has the least shaded area?**

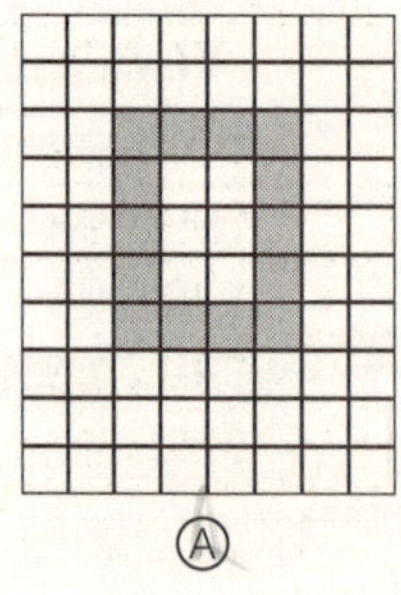
Ⓐ

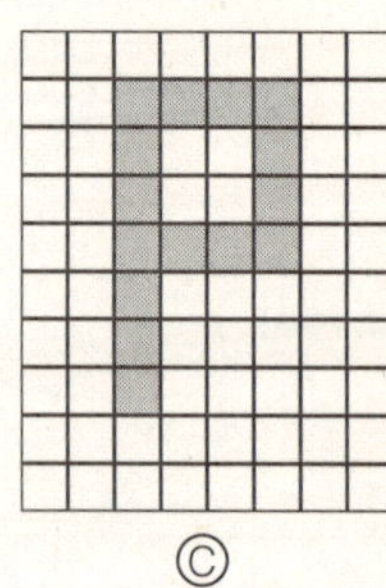
Ⓒ

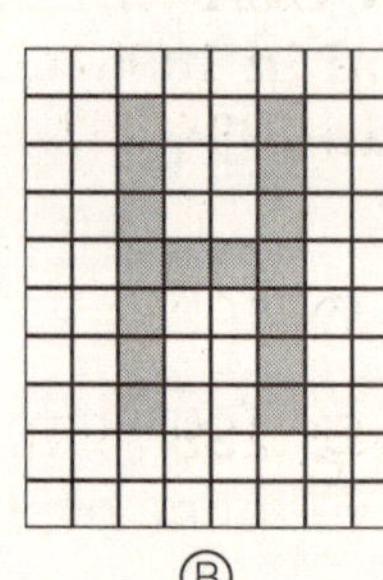
Ⓑ

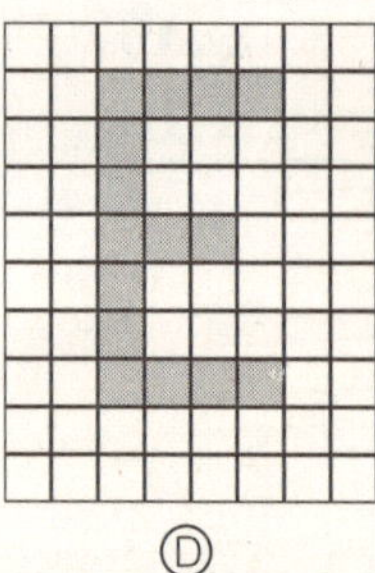
Ⓓ

Name ______________________________

BASKETBALL GAMES WON BY UCLA (1991 - 1996)

GAMES WON	YEAR
28	1991-92
22	1992-93
21	1993-94
31	1994-95
23	1995-96

In which year did UCLA win more games than they did in 1995-96, but fewer games than in 1994-95?

Ⓐ 1992-93 Ⓒ 1991-92

Ⓑ 1993-94 Ⓓ 1994-95

2 **What would the plane figure below look like if it was turned upside down?**

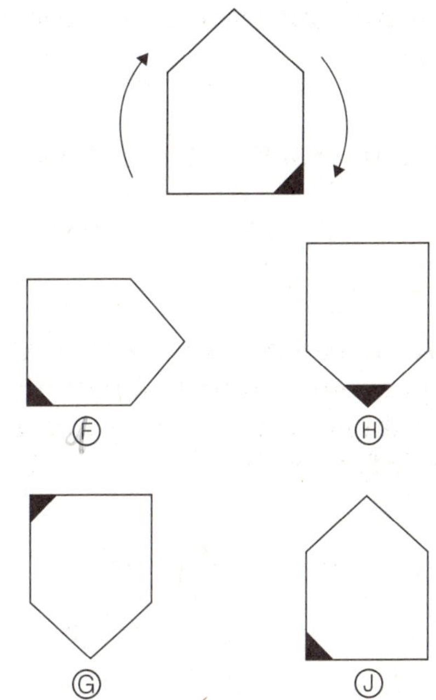

Name ______________________________

❶ $8 \times 8 =$

Ⓐ 16
Ⓑ 0
Ⓒ 64
Ⓓ 68

❷ **Mark saved $2.45 one month and $3.50 the next month. How much did he save in the two months?**

Ⓕ $ .95
Ⓖ $5.95
Ⓗ $1.15
Ⓙ $6.00

❸ **Using each of the numerals 2, 4, 5, and 6 one time only, what is the largest number you can make:**

with 6 in the ones place? __________
with 6 in the tens place? __________
with 6 in the hundreds place? __________
with 6 in the thousands place? __________

❹ **Using the numbers you wrote, find the difference between the number of greatest value and the number of least value.**

____

Name ______________________________

1 **Use the space below to draw as many ways as you can to cut this birthday cake into 12 equal pieces.**

2 **Is there any one way to cut the cake into 12 <u>equal</u> pieces so that you will get a bigger piece? Why or why not?**

____

____

____

____

Name ________________________________

How many small blocks are in this solid shape?

Ⓐ 14 blocks
Ⓒ 16 blocks
Ⓑ 18 blocks
Ⓓ 20 blocks

4 + ______ = 6 + 4

Ⓕ 10
Ⓗ 2
Ⓖ 4
Ⓙ 6

$$\begin{array}{r} \mathbf{506} \\ \mathbf{+\ 346} \\ \hline \end{array}$$

Ⓐ 942
Ⓑ 852
Ⓒ 952
Ⓓ 842
Ⓔ none of these

❹

$$\begin{array}{r} \mathbf{25} \\ \times\ \mathbf{2} \\ \hline \end{array}$$

Ⓕ 27
Ⓖ 50
Ⓗ 225
Ⓙ 40
Ⓚ none of these

Name ______________________________

1

$$\begin{array}{r} 434 \\ -\ 109 \\ \hline \end{array}$$

Ⓐ 325

Ⓑ 335

Ⓒ 235

Ⓓ 543

2 **What time is shown?**

Ⓕ 2:04

Ⓖ 4:10

Ⓗ 4:12

● 2:20

3 **Ruben was 38 years old when he moved to the United States. If Ruben was born in 1953, in what year did he move to the United States?**

Ⓐ 1981

Ⓑ 1993

Ⓒ 1991

Ⓓ 1985

4 **Read the problem below. Estimate your answer by rounding. Darren has 18 toy dinosaurs and Carmen has 23. Together they have about:**

Ⓕ 20 toy dinosaurs

Ⓖ 40 toy dinosaurs

Ⓗ 30 toy dinosaurs

● 50 toy dinosaurs

Name ______________________

1 **Mark the number that has a 6 in the thousands place.**

Ⓐ 1465
Ⓑ 6501
Ⓒ 8623
Ⓓ 4326

2 **Which number below has the greatest value?**

Ⓕ 202
Ⓖ 102
Ⓗ 211
Ⓙ 112

3

5000 + 20 + 0 =

Ⓐ 5200
Ⓑ 5020
Ⓒ 520
Ⓓ 52

4 **Find the average of these numbers.**

(2, 4)

Ⓕ 6
Ⓖ 2
Ⓗ 4
Ⓙ 3

Name ______________________________

1 **2527 rounded to the nearest hundred is:**

Ⓐ 2550　　Ⓒ 3000

Ⓑ 2600　　Ⓓ 2500

2

4 + 5 = 9

5 + 4 = 9

9 − 5 = 4

Which goes with the three above?

Ⓕ 9 + 3 = 12

Ⓖ 15 − 6 = 9

Ⓗ 5 − 4 = 1

Ⓙ 9 − 4 = 5

3 **Which fraction below names the shaded part?**

Ⓐ $\frac{1}{2}$　　Ⓒ $\frac{3}{4}$

Ⓑ $\frac{3}{6}$　　Ⓓ $\frac{1}{4}$

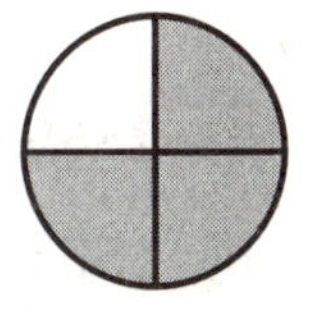

4 **Find the average of these numbers.**

(3, 5)

Ⓕ 8　　Ⓗ 4

Ⓖ 15　　Ⓙ 2

Name ___________________________

1 **The Taco House sold 283 tacos on Monday and 487 tacos on Tuesday. How many tacos were sold in all?**

Ⓐ 204 tacos
Ⓑ 690 tacos
Ⓒ 770 tacos
Ⓓ 670 tacos

2 **The temperature in degrees Fahrenheit (°F) at which water freezes is shown on the thermometer. What temperature is this?**

40
30
20

Ⓕ 31°F
Ⓖ 32°F
Ⓗ 33°F
Ⓙ 34°F

3

$$\begin{array}{r} \$\ 6.54 \\ -\ \ 5.48 \\ \hline \end{array}$$

Ⓐ $11.92
Ⓑ $1.06
Ⓒ $1.14
Ⓓ $ .16

4 **Which is the missing number?**

12, 18, 24, _____

Ⓕ 26
Ⓖ 28
Ⓗ 30
Ⓙ 32

Name ______________________________

1. **Shade a part of the figure to show what $\frac{2}{8}$ means.**

2. **Shade a part of the figure to show what $\frac{1}{4}$ means.**

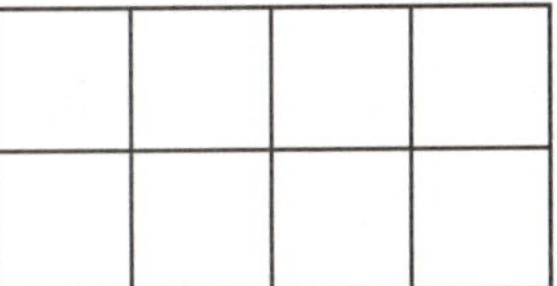

3. **Shade a part of the figure to show what $\frac{2}{8} + \frac{1}{4}$ means.**

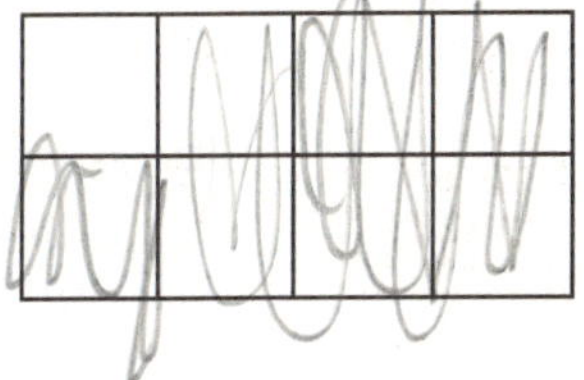

4. **Use your picture from problem 3 to write the fraction that is equal to $\frac{2}{8} + \frac{1}{4}$.**

Name ______________________________

1

$$\begin{array}{r} 36 \\ \times \quad 2 \\ \hline \end{array}$$

Ⓐ 72

Ⓑ 38

Ⓒ 62

Ⓓ 236

Ⓔ none of these

2

The drawing above is a:

Ⓕ right angle

Ⓖ ray

Ⓗ line

Ⓙ line segment

3 **Stacie played softball for six years in a row. Each year Stacie counted the number of hits she had and marked them on the graph below.**

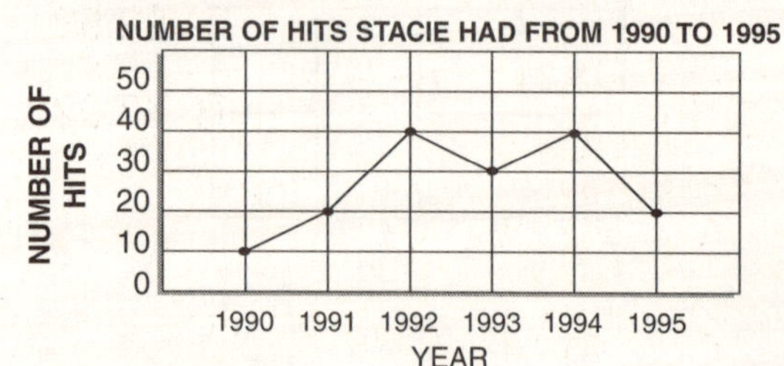

Which one of the following is true about the number of hits Stacie had between 1990 and 1995?

Ⓐ increased between 1994 and 1995

Ⓑ twice as many in 1994 as in 1991

Ⓒ stayed the same

Ⓓ decreased between 1990 and 1992

Name ______________________________

$\begin{array}{r} \$\ 6.73 \\ +\ 1.19 \\ \hline \end{array}$

Ⓐ $8.92

Ⓑ $7.82

Ⓒ $8.93

Ⓓ $7.92

2 **What shape would come next in the pattern?**

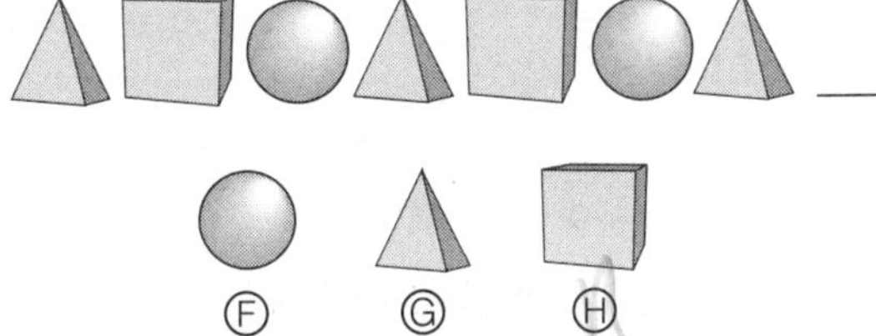

Ⓕ Ⓖ Ⓗ

3 **Which is the numeral that is one thousand forty-three?**

Ⓐ 1043

Ⓑ 1430

Ⓒ 1403

Ⓓ 1340

4 **Which figure is a hexagon?**

Ⓕ 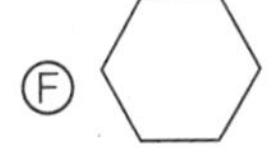

Ⓗ 

Ⓖ 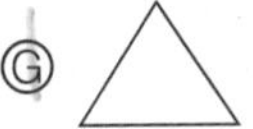

Ⓙ

Name ____________________

1 **Pat is fourth in line. How many are before her?**

Ⓐ 3 Ⓒ 5

Ⓑ 4 Ⓓ 2

2 **Read the problem below. Estimate your answer by rounding. Estella weighs 46 pounds. Mike weighs 52 pounds. Together they weigh about:**

Ⓕ 50 pounds Ⓗ 200 pounds

Ⓖ 100 pounds Ⓙ 300 pounds

3 **Using each of the numerals 5, 3, 2 and 1 one time only, what is the largest number you can make:**

with 1 in the ones place? ________

with 1 in the tens place? ________

with 1 in the hundreds place? ________

with 1 in the thousands place? ________

4 **Using the numbers you wrote, find the difference between the number of greatest value and the number of least value.**

____

Name ______________________________

1

600 + 388 + 6 =

Ⓐ 982

Ⓑ 974

Ⓒ 994

Ⓓ 1094

2 **John needs to jog 15 miles to win an award. He has jogged 9 miles. How many more miles must John jog?**

Ⓕ 6 miles

Ⓗ 7 miles

Ⓖ 24 miles

Ⓙ 8 miles

3 **Find the average of these numbers:**

(3, 7)

Ⓐ 5

Ⓒ 10

Ⓑ 2

Ⓓ 21

4

$$2\overline{)24}$$

Ⓕ 11

Ⓗ 12

Ⓖ 21

Ⓙ 10

Name ______________________

$$\begin{array}{r} \$\ 8.68 \\ -\ \ 6.59 \\ \hline \end{array}$$

Ⓐ $2.11

Ⓑ $1.09

Ⓒ $2.01

Ⓓ $2.09

Ⓔ none of these

6 + 4 = 10

4 + 6 = 10

10 − 6 = 4

Which goes with the three above?

Ⓕ 10 + 6 = 16

Ⓖ 4 + 4 = 8

Ⓗ 10 − 4 = 6

Ⓙ 6 − 4 = 2

3

$$\begin{array}{r} 45 \\ \times\ \ 2 \\ \hline \end{array}$$

Ⓐ 245

Ⓑ 900

Ⓒ 80

Ⓓ 47

Ⓔ none of these

4 **Mark the odd number.**

Ⓕ 43

Ⓖ 12

Ⓗ 20

Ⓙ 30

Name ______________________________

Look at the different parts of the Venn diagram below and answer the questions that follow.

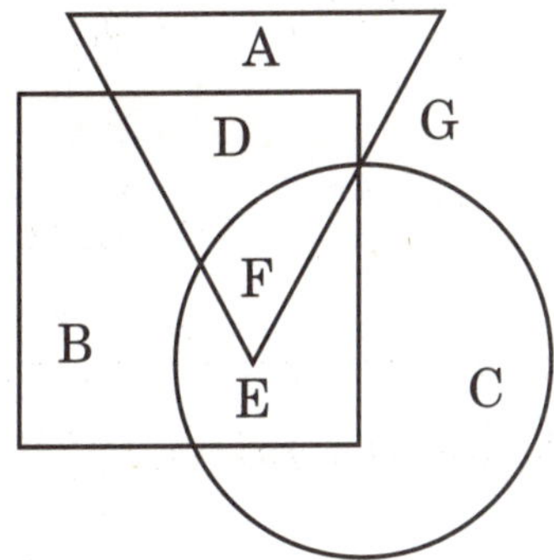

1 **What letter is in the triangle but not in the square or circle?**

Ⓐ A　　Ⓒ F

Ⓑ D　　Ⓓ G

2 **What letter is in the circle and square but not in the triangle?**

Ⓕ B　　Ⓗ E

Ⓖ F　　Ⓙ C

3 **What letter is in the circle, the square, and the triangle?**

Ⓐ A　　Ⓒ F

Ⓑ D　　Ⓓ G

4 **What is the letter that does not belong to any of the three figures?**

Ⓕ C　　Ⓗ A

Ⓖ G　　Ⓙ B

Name ____________________

1 **743 rounded to the nearest hundred is:**

Ⓐ 750 Ⓒ 700

Ⓑ 800 Ⓓ 600

2

6						
5		■				
4					●	
3						
2						
1			▲			
0	1	2	3	4	5	6

Look at the graph. Where is the ▲?

Ⓕ (4,2)

Ⓖ (3,1)

Ⓗ (5,4)

Ⓙ (1,4)

3 $3\overline{)36}$

Ⓐ 13 Ⓒ 9

Ⓑ 12 Ⓓ 10

4 **Today is Saturday the 21st of June. What is the day of the week on the 29th of June?**

Ⓕ Saturday Ⓗ Monday

Ⓖ Sunday Ⓙ Tuesday

Name ______________________________

1

$(4 + 7) + 8 =$ _____ $+ (7 + 8)$

Ⓐ 11
Ⓑ 4
Ⓒ 19
Ⓓ 15

2

$(8 \times 1000) + (9 \times 1) =$

Ⓕ 80,009
Ⓖ 8009
Ⓗ 8090
Ⓙ 8900

3 **Lauren has five coins in her purse that total 45¢. What is one possible combination of five coins that she might have that equals 45¢?**

4 **Find another combination of five coins that also equals 45¢.**

Name ______________________________

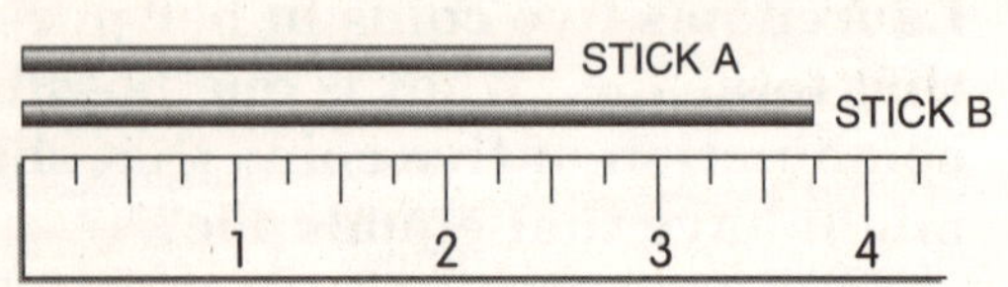

1 **How long is stick A?**

Ⓐ 2 inches

Ⓑ $2\frac{1}{4}$ inches

Ⓒ $2\frac{1}{2}$ inches

Ⓓ $2\frac{3}{4}$ inches

2 **How much shorter is stick A than stick B?**

Ⓕ 1 inch

Ⓖ $1\frac{1}{4}$ inches

Ⓗ $1\frac{1}{2}$ inches

Ⓙ $2\frac{1}{2}$ inches

3 **Which symbol goes in the circle below?**

$$15 - 5 \bigcirc 2 \times 10$$

Ⓐ >

Ⓑ <

Ⓒ =

4

$$\begin{array}{r} \$\ 9.33 \\ -\ 4.67 \\ \hline \end{array}$$

Ⓕ \$5.34

Ⓖ \$4.76

Ⓗ \$5.66

Ⓙ \$4.66

Ⓚ none of these

Name ______________________________

1 **There are 12 eggs in a dozen. How many eggs are there in 7 dozen?**

Ⓐ 60 eggs
Ⓑ 72 eggs
Ⓒ 84 eggs
Ⓓ 96 eggs

2 **There are 11 cages in the zoo. There are 3 monkeys in each cage. How many monkeys are in the zoo?**

Ⓕ 8 monkeys
Ⓖ 14 monkeys
Ⓗ 113 monkeys
Ⓙ 33 monkeys

3 **What fraction of all the circles is shaded?**

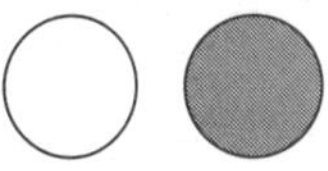

Ⓐ $\frac{0}{1}$ Ⓑ $\frac{1}{1}$ Ⓒ $\frac{1}{2}$ Ⓓ $\frac{2}{1}$

4 **1850 rounded to the nearest thousand is:**

Ⓕ 1000
Ⓖ 2000
Ⓗ 1800
Ⓙ 1900

Name ______________________

1 **In 3 baseball games the Angels scored 3, 2, and 7 runs. What was the Angels' average score?**

Ⓐ 5 runs Ⓒ 12 runs

Ⓑ 9 runs Ⓓ 4 runs

2

$$3\overline{)69}$$

Ⓕ 33 Ⓗ 13

Ⓖ 23 Ⓙ 21

3

$$4 \times \square = 12$$

Ⓐ 3 Ⓒ 48

Ⓑ 8 Ⓓ 6

4

$$\begin{array}{r} 260 \\ -\ 114 \\ \hline \end{array}$$

Ⓕ 374

Ⓖ 154

Ⓗ 146

Ⓙ 150

Name ______________________________

1

$$\begin{array}{r} \$\ 5.56 \\ -\ 3.97 \\ \hline \end{array}$$

Ⓐ \$2.41

Ⓑ \$1.59

Ⓒ \$2.59

Ⓓ \$1.69

Ⓔ none of these

2

3 + 2 =

Ⓕ 5 + 2

Ⓖ 5 − 2

Ⓗ 2 + 3

Ⓙ 2 × 3

3 **Which is the number with the least value?**

Ⓐ 404

Ⓑ 414

Ⓒ 444

Ⓓ 440

4

$$\begin{array}{r} 105 \\ \times\ \ 2 \\ \hline \end{array}$$

Ⓕ 112

Ⓖ 210

Ⓗ 122

Ⓙ 107

Ⓚ none of these

Name ________________________________

1 **There were 64 pieces of bubble gum in a bubble gum machine. For the past 4 days, students have taken out 13 pieces each day. How many pieces of bubble gum are left?**

Ⓐ 51 pieces
Ⓑ 22 pieces
Ⓒ 12 pieces
Ⓓ 14 pieces

2 **Which list has the numbers in order from greatest value to least value?**

Ⓕ $\frac{1}{4}$, $\frac{1}{3}$, $\frac{1}{2}$
Ⓖ $\frac{1}{3}$, $\frac{1}{4}$, $\frac{1}{2}$
Ⓗ $\frac{1}{2}$, $\frac{1}{4}$, $\frac{1}{3}$
Ⓙ $\frac{1}{2}$, $\frac{1}{3}$, $\frac{1}{4}$

3 **Use an inch ruler and this diagram to help you answer the question. How many inches are used to show the distance along the path from home to school to work?**

Home
School
Work

Ⓐ 3 inches
Ⓑ $3\frac{1}{2}$ inches
Ⓒ 4 inches
Ⓓ $4\frac{1}{2}$ inches

Name ______________________________

1 **Which list has the numbers in order from least value to greatest value?**

Ⓐ $\frac{1}{2}, \frac{2}{3}, \frac{3}{4}$

Ⓒ $\frac{3}{4}, \frac{1}{2}, \frac{2}{3}$

Ⓑ $\frac{3}{4}, \frac{2}{3}, \frac{1}{2}$

Ⓓ $\frac{1}{2}, \frac{3}{4}, \frac{2}{3}$

2 **Janet took a trip. She drove 386 miles on Monday, 517 miles on Tuesday, and 410 miles on Wednesday. *About* how many miles did she drive in all?**

Ⓕ 1400 miles

Ⓗ 1300 miles

Ⓖ 1500 miles

Ⓙ 1200 miles

3 **HOME RUNS HIT BY ROGER MARIS (1959 - 1963)**

HOME RUNS	YEAR
16	1959
40	1960
61	1961
33	1962
23	1963

In which year did Roger Maris hit more home runs than he did in 1962, but fewer home runs than in 1961?

Ⓐ 1959

Ⓒ 1961

Ⓑ 1960

Ⓓ 1963

Name ______________________________

1 **Alex bought some candy at the store. It cost $1.00. He gave the clerk the coins shown below.**

How much more does Alex owe?

Ⓐ 9¢ Ⓒ 19¢

Ⓑ 4¢ Ⓓ 14¢

2 **Which list has the numbers in order from greatest value to least value?**

Ⓕ $2\frac{1}{2}, 1\frac{2}{3}, 1\frac{1}{4}, 2\frac{1}{3}$

Ⓖ $2\frac{1}{2}, 2\frac{1}{3}, 1\frac{2}{3}, 1\frac{1}{4}$

Ⓗ $1\frac{1}{4}, 1\frac{2}{3}, 2\frac{1}{3}, 2\frac{1}{2}$

Ⓙ $2\frac{1}{3}, 2\frac{1}{2}, 1\frac{1}{4}, 1\frac{2}{3}$

3) 7

Ⓐ 2 R1 Ⓒ 20 R1

Ⓑ 3 RI Ⓓ 2

Name ______________________________

1 **Look at this line segment.**

Which line segment below appears to be twice as long as the line segment above?

Ⓐ

Ⓑ

Ⓒ

Ⓓ

2

$4\overline{)9}$

Ⓕ 2 R2

Ⓖ 2 R1

Ⓗ 20

Ⓙ 21

3 **Sonia bought 3 bags of peanuts for 75¢ a bag. How much change should she receive from a $5 bill?**

Ⓐ $3.50

Ⓑ $3.00

Ⓒ $2.75

Ⓓ $2.25

Name ______________________________

1

$4\overline{)48}$

Ⓐ 18
Ⓑ 14
Ⓒ 24
Ⓓ 12

2 **5009 is read as:**

Ⓕ five hundred nine
Ⓖ fifty thousand nine
Ⓗ five thousand nine
Ⓙ five thousand ninety

3 **Christmas is in December. Thanksgiving is in the month before December. Thanksgiving is in what month?**

Ⓐ January
Ⓑ November
Ⓒ March
Ⓓ September

4 **Ed is ninth in line. How many are <u>before</u> him?**

Ⓕ 8
Ⓖ 7
Ⓗ 10
Ⓙ 0

Name ______________________________

1

$3\overline{)96}$

Ⓐ 3 R2
Ⓑ 302
Ⓒ 32
Ⓓ 28

2 **Read the problem below. Estimate your answer by rounding. Katie drove 133 miles one day and 171 the next day. In the two days, Katie drove about:**

Ⓕ 200 miles
Ⓖ 300 miles
Ⓗ 250 miles
Ⓙ 350 miles

3 **There are 18 holes on a golf course. How many holes are there on 6 golf courses?**

Ⓐ 102 holes
Ⓑ 108 holes
Ⓒ 114 holes
Ⓓ 96 holes

4 **If a car travels at an average speed of 28 miles per hour, how far can the car go in 5 hours?**

Ⓕ 280 miles
Ⓖ 120 miles
Ⓗ 140 miles
Ⓙ 5.6 miles

Name ______________________________

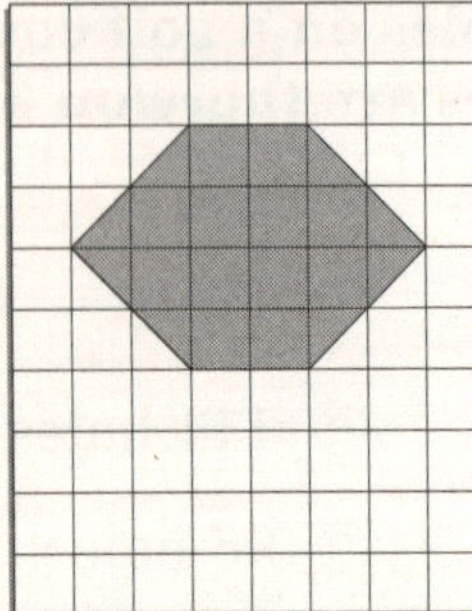

The shaded figure on the grid to the left has an area of about:

Ⓐ 9 square units.

Ⓑ 12 square units.

Ⓒ 16 square units.

Ⓓ 18 square units.

Ⓔ none of these

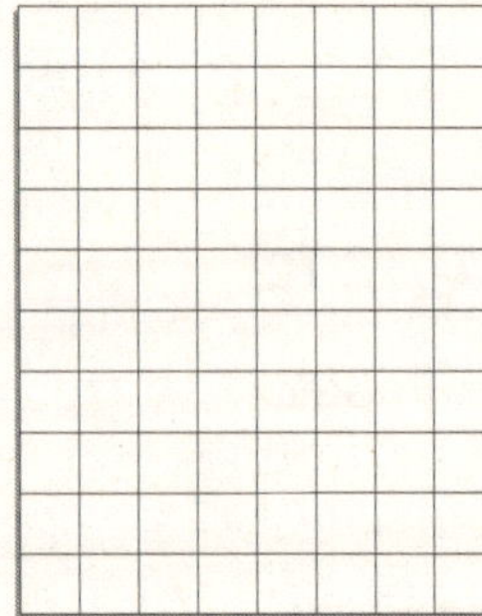

Draw a rectangle on the blank grid to the right that has the same area.

2

$$\begin{array}{r} 370 \\ -\ 129 \\ \hline \end{array}$$

Ⓕ 241

Ⓖ 259

Ⓗ 251

Ⓙ 250

Ⓚ none of these

3

$$\begin{array}{r} 415 \\ \times\ \ 2 \\ \hline \end{array}$$

Ⓐ 830

Ⓑ 820

Ⓒ 840

Ⓓ 930

Ⓔ none of these

Name ______________________________

600 =

Ⓐ _6_ hundreds _6_ tens _6_ ones

Ⓑ _6_ hundreds _0_ tens _0_ ones

Ⓒ _0_ hundreds _6_ tens _0_ ones

Ⓓ _6_ hundreds _0_ tens _6_ ones

2

$$\begin{array}{r} \$\ 2.17 \\ +\ 5.39 \\ \hline \end{array}$$

Ⓕ $8.56

Ⓖ $7.56

Ⓗ $7.46

Ⓙ $8.46

3 **Nina has 48 cows on her father's farm. She wants to put the same number of cows in each of their 4 stalls. How many cows should she put in each stall?**

Ⓐ 24 cows

Ⓑ 12 cows

Ⓒ 44 cows

Ⓓ 10 cows

4 **Round 6315 to the nearest thousand.**

Ⓕ 6300

Ⓖ 6000

Ⓗ 6500

Ⓙ 7000

Name ______________________________

1 **Greg, Amy, Jeff, and Denise are all friends. Each has a different colored bike. The colors are blue, green, red, and purple. Greg does not have a red or purple bike. Amy does not have a blue or red bike. Jeff has a blue bike. Who has the purple bike? You may want to use the table below to help you find the answer.**

	BLUE	GREEN	RED	PURPLE
GREG				
AMY				
JEFF				
DENISE				

__________ has the purple bike.

Name ______________________________

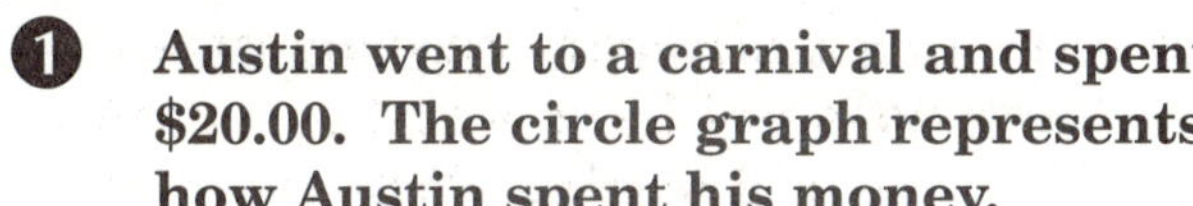

❶ **Austin went to a carnival and spent $20.00. The circle graph represents how Austin spent his money.**

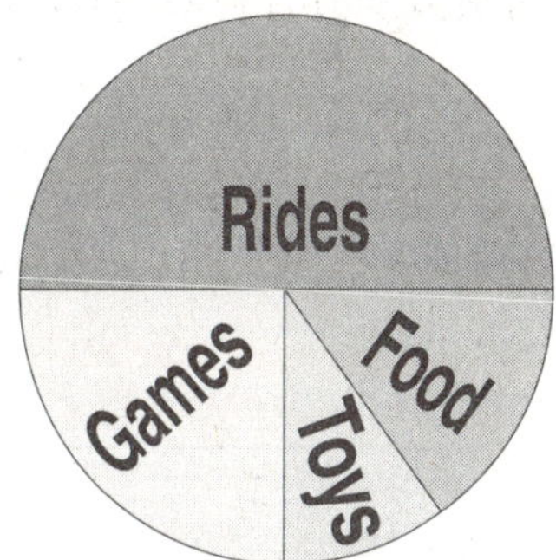

About how much money did Austin spend on:

rides __________ food __________

games __________ toys __________

❷ **Make a list of how you would spend $20.00 at a carnival. Your list should be different than Austin's. Then draw any type of graph that shows how you would spend the $20.00.**

Name ______________________________

1

$(7 \times 1000) + (5 \times 10) + (4 \times 1) =$

Ⓐ 7504

Ⓑ 7540

Ⓒ 7054

Ⓓ 754

2

$$\begin{array}{r} 460 \\ -\ 237 \\ \hline \end{array}$$

Ⓕ 233

Ⓖ 223

Ⓗ 237

Ⓙ 230

3

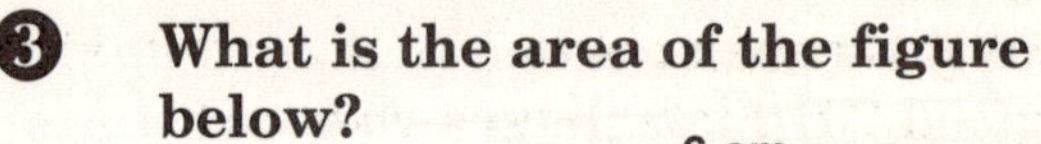

What is the area of the figure below?

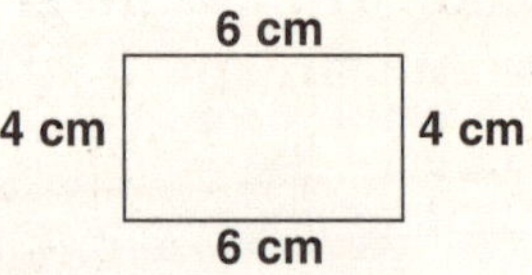

Ⓐ 10 sq. cm

Ⓑ 20 sq. cm

Ⓒ 32 sq. cm

Ⓓ 24 sq. cm

4

4 birds in a tree. Some birds flew away. How many birds are left? What else do you need to know to answer the question?

Ⓕ Where was the tree?

Ⓖ How many birds could fly?

Ⓗ How many birds flew away?

Name ____

❶ **A small pizza costs $7. How much would 20 small pizzas cost?**

Ⓐ $14

Ⓑ $140

Ⓒ $210

Ⓓ $104

❷

$$\begin{array}{r} \$\ \ 4.68 \\ +\ \ \ 3.37 \\ \hline \end{array}$$

Ⓕ $8.05

Ⓖ $1.31

Ⓗ $8.15

Ⓙ $8.95

Ⓚ none of these

❸ **Which figure shows a diagonal line?**

Ⓐ

Ⓑ

Ⓒ

Ⓓ

❹ **What is the number that tells how many days in 3 weeks?**

Ⓕ 14

Ⓖ 15

Ⓗ 21

Ⓙ 28

Name ______________________________

1 **Don rode his bicycle for 3 hours traveling about 19 miles per hour. About how many miles did he travel in the 3 hours?**

Ⓐ 50 miles Ⓒ 70 miles

Ⓑ 60 miles Ⓓ 80 miles

---

2 **Which list has the numbers in order from least value to greatest value?**

Ⓕ $1\frac{1}{8}, 1\frac{2}{3}, 2\frac{1}{4}, 2\frac{1}{2}$

Ⓖ $1\frac{2}{3}, 1\frac{1}{8}, 2\frac{1}{4}, 2\frac{1}{2}$

Ⓗ $2\frac{1}{2}, 2\frac{1}{4}, 1\frac{2}{3}, 1\frac{1}{8}$

Ⓙ $1\frac{1}{8}, 1\frac{2}{3}, 2\frac{1}{2}, 2\frac{1}{4}$

3 **What would the figure below look like if it was turned upside down?**

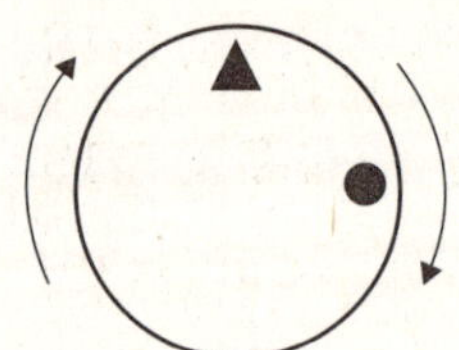

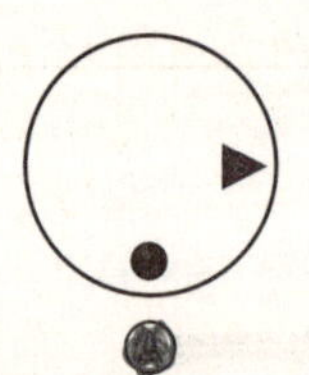

Ⓐ

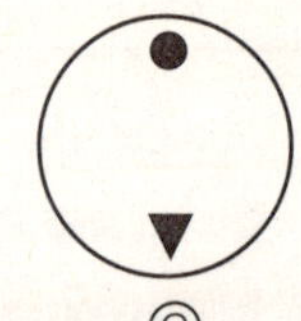

Ⓒ

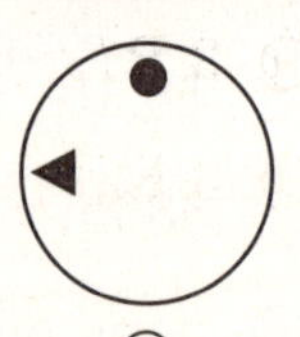

Ⓑ

Ⓓ

Name ______________________________

1 **Which list has the numbers in order from greatest value to least value?**

Ⓐ 5.41, 5.46, 5.23, 5.19

Ⓑ 5.46, 5.41, 5.23, 5.19

Ⓒ 5.46, 5.41, 5.19, 5.23

Ⓓ 5.19, 5.23, 5.46, 5.41

2

$2\overline{)7}$

Ⓕ 25

Ⓖ 4

Ⓗ 3 R1

Ⓙ 3 R2

3 **Use an inch ruler and this diagram to help you answer the question. How many inches are used to show the distance along the path from Home to the Park, to the Theater, to the Store, and back Home again?**

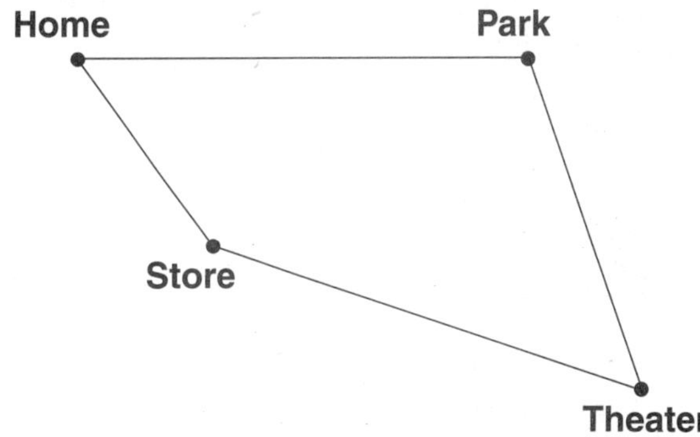

Ⓐ $5\frac{1}{2}$ inches

Ⓑ 6 inches

Ⓒ $6\frac{1}{2}$ inches

Ⓓ $7\frac{1}{2}$ inches

Name ______________________________

1 **Two girls want to share 12 marbles equally. How many marbles will each girl get?**

Ⓐ 24 marbles
Ⓑ 8 marbles
Ⓒ 6 marbles
Ⓓ 4 marbles

2

$$\begin{array}{r} 215 \\ \times \quad 4 \\ \hline \end{array}$$

Ⓕ 840
Ⓖ 842
Ⓗ 219
Ⓙ 860

3 **13 − 4 = 9**
13 − 9 = 4
4 + 9 = 13

Which goes with the three above?

Ⓐ 13 − 4 = 4
Ⓑ 9 + 4 = 13
Ⓒ 13 + 4 = 17
Ⓓ 9 − 4 = 5

4 **Mr. Dillard has 25 feet of wire fencing material. Which of the gardens below can he fence with the material he has?**

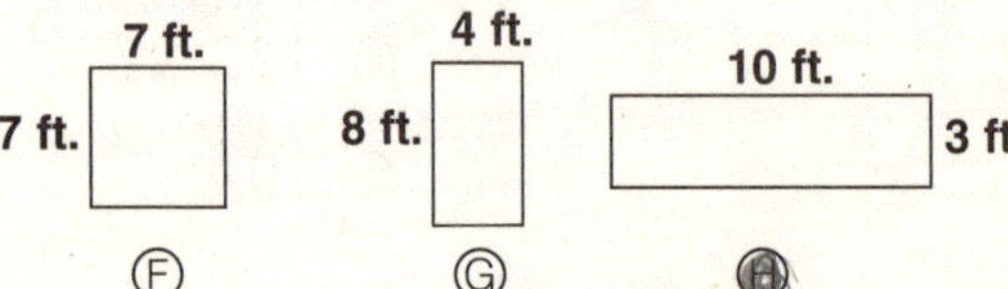

Name ______________________________

1 **How many inches are in 2 feet?**

Ⓐ 32 inches
Ⓑ 20 inches
Ⓒ 24 inches
Ⓓ 8 inches

2

$$2\overline{)248}$$

Ⓕ 1024
Ⓖ 124
Ⓗ 122 R2
Ⓙ 104

3 **A grocer is stacking cans on a shelf. He puts 15 cans in the first row, 12 cans in the second row, and 9 in the third row. If this pattern continues, how many cans will be in the <u>fifth</u> row?**

Ⓐ 36 cans
Ⓑ 6 cans
Ⓒ 3 cans
Ⓓ 1 can

Explain your answer in words or a picture.

Name ______________________________

The Venn diagram below shows what a group of high school students think about cars and trucks.

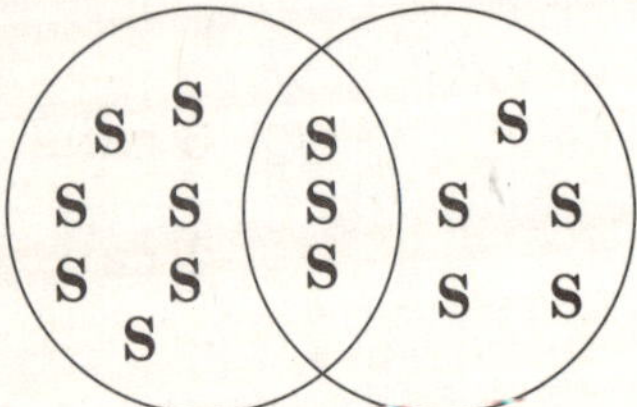

S = 1 Student

❶ **How many students like cars only?**

Ⓐ 3 students
Ⓑ 7 students
Ⓒ 5 students
Ⓓ 10 students

❷ **How many students like both cars and trucks?**

Ⓕ 3 students
Ⓖ 12 students
Ⓗ 15 students
Ⓙ 10 students

❸ **9 + 7 =**

Ⓐ 15
Ⓑ 16
Ⓒ 17
Ⓓ 18

❹

$$\begin{array}{r} 305 \\ \times \quad 3 \\ \hline \end{array}$$

Ⓕ 915
Ⓖ 1015
Ⓗ 905
Ⓙ 308

Name ______________________________

1 **Which numeral names one thousand, two hundred, forty?**

Ⓐ 10,240 Ⓒ 1,240

Ⓑ 12,400 Ⓓ 2,040

2 **What is $15.75 rounded to the nearest dollar?**

Ⓕ $15.00 Ⓗ $16.00

Ⓖ $15.50 Ⓙ $20.00

3 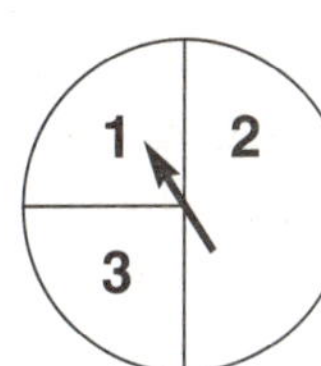

On one spin of the spinner, which number has the highest probability of being selected?

Ⓐ 1 Ⓒ 2

Ⓑ 3 Ⓓ 4

4 **Roy's grandmother is 68 years old. Roy is 9 years old. How many years older than Roy is his grandmother?**

Ⓕ 59 years Ⓗ 58 years

Ⓖ 77 years Ⓙ 49 years

Name ____________________

1 **Suppose you want to estimate the distance between your home in city A and your grandmother's home in city B. Which one of the units of measure below would be best to use?**

Ⓐ inches
Ⓑ feet
Ⓒ yards
● miles

3 **5067 is read as**

Ⓐ five thousand, six hundred seven
Ⓑ five thousand, sixty-seven
● five hundred sixty-seven
Ⓓ fifty sixty-seven
Ⓔ none of these

2 **3 quarters + 1 nickel =**

Ⓕ 65¢
Ⓖ 50¢
Ⓗ 80¢
● 85¢

4 **Glenn, Scott, and Juan all played on the same soccer team last year. Scott scored 5 goals more than Glenn. Juan scored 2 goals less than Scott. How many more goals did Juan score than Glenn?**

● 5 goals
Ⓖ 2 goals
Ⓗ 7 goals
Ⓙ 3 goals

Name ______________________________

Cody has nine coins in his pocket. The sum of the nine coins is 81¢. What is one possible combination of nine coins that Cody might have in his pocket?

Find other combinations of nine coins that also equal 81¢.

Name ________________________________

1

$\square \times 6 = 42$

Ⓐ 9
Ⓑ 7
Ⓒ 6
Ⓓ 8

---

2 **A movie began at 1:00 P.M. and was over at 1:45 P.M. How many minutes long was the movie?**

Ⓕ 15 minutes
Ⓖ 65 minutes
Ⓗ 45 minutes
Ⓙ 100 minutes

3 **Three friends went to a movie together and each had a different amount of money to spend on food as shown in the graph below.**

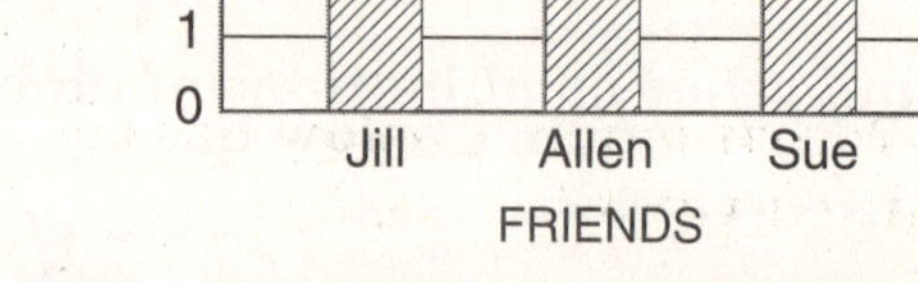

If they decided to put all of their money together and share it equally, how much money would each have?

Ⓐ $9
Ⓑ $4
Ⓒ $2
Ⓓ $3

Name ____________________

1

$$4\overline{)480}$$

Ⓐ 110
Ⓑ 120
Ⓒ 1020
Ⓓ 12 R0

2 **Which number below has the least value?**

Ⓕ 1201
Ⓖ 1021
Ⓗ 2101
Ⓙ 2021

3

$$\begin{array}{r} 950 \\ -\ 635 \\ \hline \end{array}$$

Ⓐ 315
Ⓑ 320
Ⓒ 325
Ⓓ 1585

4 **Eight girls and twelve boys were playing in the park. Six had to go home. How many were left to play?**

Ⓕ 6
Ⓖ 14
Ⓗ 16
Ⓙ 10

Name ______________________________

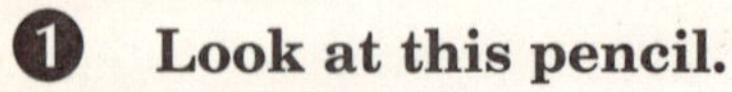

1 **Look at this pencil.**

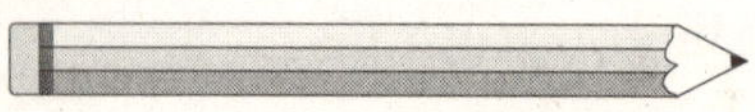

Which pencil below appears to be half as long as the pencil above?

Ⓐ 

Ⓑ 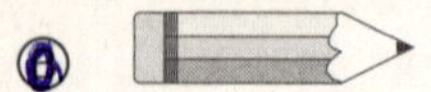

Ⓒ 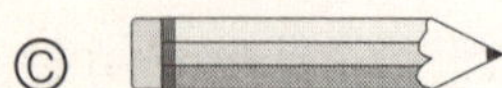

Ⓓ 

2 **Suppose you want to estimate the weight of a wooden ruler. Which one of the units of measure below would be best to use?**

Ⓕ ounces

Ⓖ pounds

Ⓗ tons

Ⓙ inches

3 **Mike drove 505 miles in one day. His car averaged 24 miles per gallon. *About* how many gallons of gas did he use on this trip?**

Ⓐ 25 gallons

Ⓑ 20 gallons

Ⓒ 15 gallons

Ⓓ 30 gallons

Name ______________________________

1 **Erica had 4 dolls. Mother bought her 3 more. How many dolls does Erica have now? Select the correct number sentence.**

Ⓐ $4 - 3 = 1$

Ⓑ $4 + 3 = 7$

Ⓒ $4 \times 3 = 12$

Ⓓ $7 - 3 = 4$

2

$3\overline{)8}$

Ⓕ 2 R1

Ⓖ 20 R2

Ⓗ 2 R2

Ⓙ 3

3 **Look at the addition problems below.**

27	+	**5**	=	☐
28	+	**8**	=	☐
31	+	**9**	=	☐

If two of the numbers are switched, then the same answer will belong in all three boxes. Which two numbers should switch places?

Ⓐ switch 5 and 8

Ⓑ switch 5 and 9

Ⓒ switch 8 and 9

Ⓓ switch 31 and 9

Name ______________________

1 **Which is the numeral 5 hundreds, 0 tens, 3 ones?**

Ⓐ 530 Ⓒ 503

Ⓑ 035 Ⓓ 305

2

$$\begin{array}{r} 151 \\ \times \quad 4 \\ \hline \end{array}$$

Ⓕ 604

Ⓖ 155

Ⓗ 504

Ⓙ 420

Ⓚ none of these

3 **Write the fraction for the part that is shaded in the circle.**

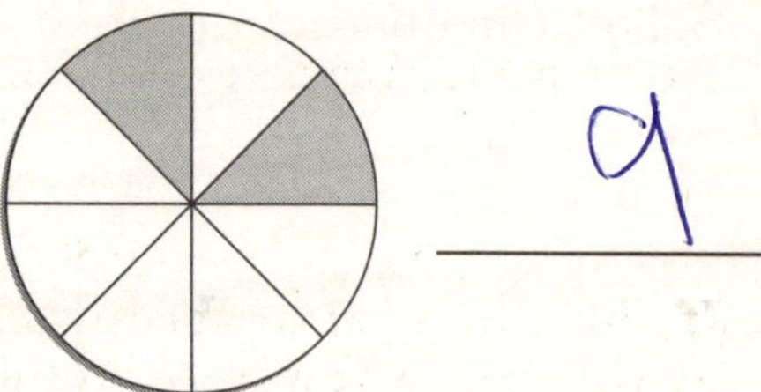

9

4 **Write an equivalent fraction to your answer in question 3.** 37 **Shade the circle below to match the equivalent fraction.**

Name ______________________________

1 **John has a strip of string cheese that is 12 inches long. He wants to cut it into equal pieces using only 2 cuts. How long would each equal strip be?**

Ⓐ 2 inches Ⓑ 3 inches Ⓒ 4 inches Ⓓ 6 inches

Draw a picture to prove that your answer is correct.

If John wanted to share the 12 inch strip of string cheese equally between 6 people, what is the minimum (fewest) number of cuts needed? ________ cuts

2

$$\begin{array}{r} \$\ 1.55 \\ +\ 6.80 \\ \hline \end{array}$$

Ⓕ $8.45

Ⓖ $8.30

Ⓗ $7.35

Ⓙ $8.35

3 **Which number names the shaded parts?**

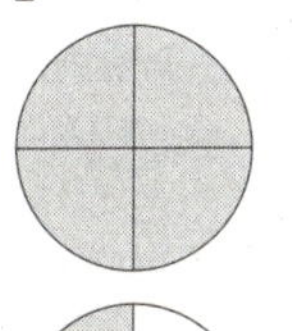

Ⓐ 6 Ⓒ $1\frac{1}{2}$

Ⓑ $2\frac{1}{2}$ Ⓓ $1\frac{2}{8}$

Name ____________________

1. **Kim picked 36 plums from her tree. She wants to divide them up equally between her 3 sisters. How many plums should she give each sister?**

Ⓐ 10 plums Ⓒ 14 plums

Ⓑ 12 plums Ⓓ 13 plums

2. **How many sides has a rectangle?**

Ⓕ 3 Ⓗ 6

Ⓖ 4 Ⓙ 8

3. **What is the average of these numbers?**

(3, 3, 6)

Ⓐ 12  9

Ⓑ 3 Ⓓ 4

4. **Which of the following figures is a sphere?**

Ⓕ 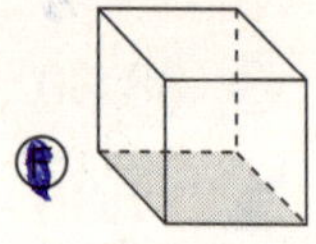 Ⓗ 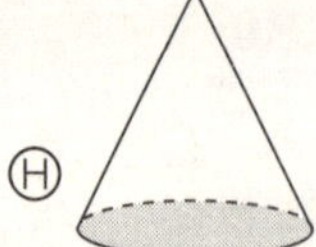

Ⓖ 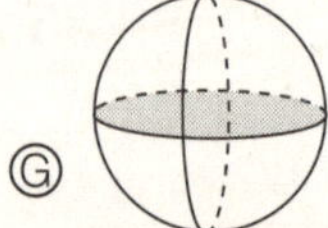 Ⓙ 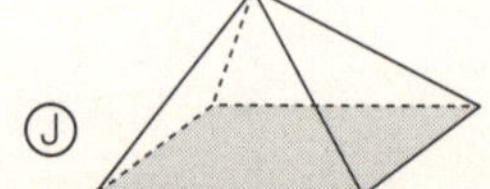

Name ______________________________

1 **Which number is divisible by 3?**

Ⓐ 25 Ⓒ 14

Ⓑ 10 Ⓓ 24

2 **(4 + 9) + 7 is the same as:**

Ⓕ (4 + 8) + 9 Ⓗ (7 + 9) + 5

Ⓖ 4 + (9 + 7) Ⓙ 4 + (7 + 16)

3 **7100 is read as:**

Ⓐ seventy-one

Ⓑ seven thousand, one hundred

Ⓒ seven one hundred

Ⓓ seven thousand one

4 **Which number is an odd number?**

Ⓕ 12 Ⓗ 93

Ⓖ 66 Ⓙ 84

Name ______________________________

1

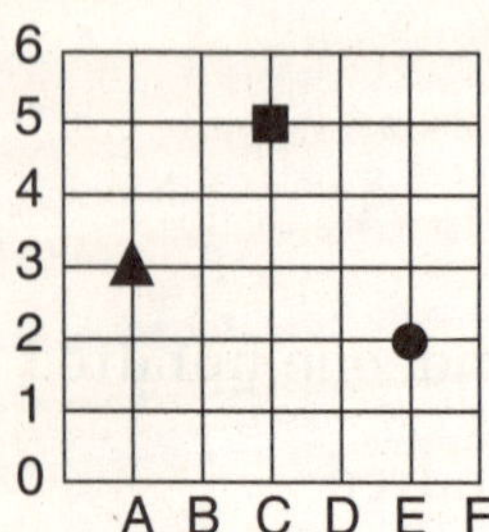

Look at the graph. What is located at (C, 5)?

Ⓐ ▲

Ⓑ ■

Ⓒ ●

2 **The line segment AB represents a:**

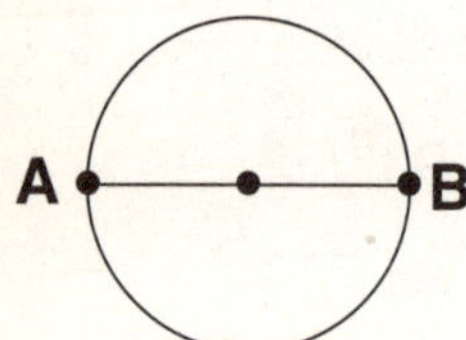

Ⓕ radius

Ⓖ perimeter

Ⓗ diagonal

Ⓙ diameter

3

$$\begin{array}{r} 143 \\ \times \quad 3 \\ \hline \end{array}$$

Ⓐ 146

Ⓑ 429

Ⓒ 439

Ⓓ 426

4 **Which is the missing number?**

25, 20, 15, ______

Ⓕ 5

Ⓖ 10

Ⓗ 60

Ⓙ 30

Name ______________________________

1 **Which number is divisible by 5?**

Ⓐ 9 Ⓒ 12

Ⓑ 15 Ⓓ 21

$$\begin{array}{r} 200 \\ \times \quad 3 \\ \hline \end{array}$$

Ⓐ 633

Ⓑ 500

Ⓒ 900

Ⓓ 203

Ⓔ none of these

2

IN	OUT
2	5
3	6
4	7
5	8

In the table on the left, the rule is:

Ⓕ subtract 2

Ⓖ add 2

Ⓗ subtract 3

Ⓙ add 3

Ⓚ none of these

4 **Rearrange the digits 1, 4, and 9 to make as many different three-digit numbers as you can.**

Name ______________________

1 **What is $57.35 rounded to the nearest dollar?**

Ⓐ $56.00 Ⓒ $58.00

Ⓑ $50.00 Ⓓ $57.00

3

$$2\overline{)220}$$

Ⓐ 1010 Ⓒ 110

Ⓑ 100 Ⓓ 100 R1

2 **In the figure below, is the dotted line a line of symmetry?**

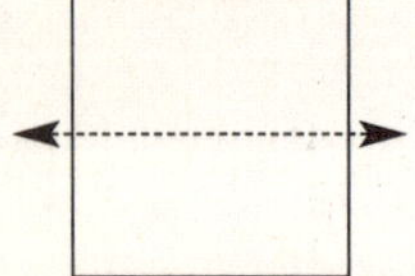

Ⓕ yes Ⓖ no

Draw all possible lines of symmetry.

4 **Which number is an even number?**

Ⓕ 45 Ⓗ 101

Ⓖ 83 Ⓙ 70

Name ____________________

$\frac{2}{5} + \frac{1}{5} =$

Ⓐ $\frac{2}{25}$

Ⓑ $\frac{3}{5}$

Ⓒ $\frac{3}{10}$

Ⓓ $\frac{1}{5}$

$9 \times 3 = 3 \times \square$

Ⓕ 12

Ⓖ 27

Ⓗ 9

Ⓙ 3

$$\begin{array}{r} 401 \\ -\ 229 \\ \hline \end{array}$$

Ⓐ 228

Ⓑ 282

Ⓒ 208

Ⓓ 172

4 **Which fraction tells how much of the figure below is shaded?**

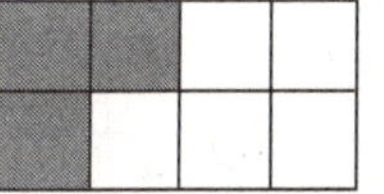

Ⓕ $\frac{3}{8}$

Ⓖ $\frac{3}{5}$

Ⓗ $\frac{5}{8}$

Ⓙ $\frac{1}{2}$

Name ______________________________

❶ **Which grid has thirty-two hundredths shaded?**

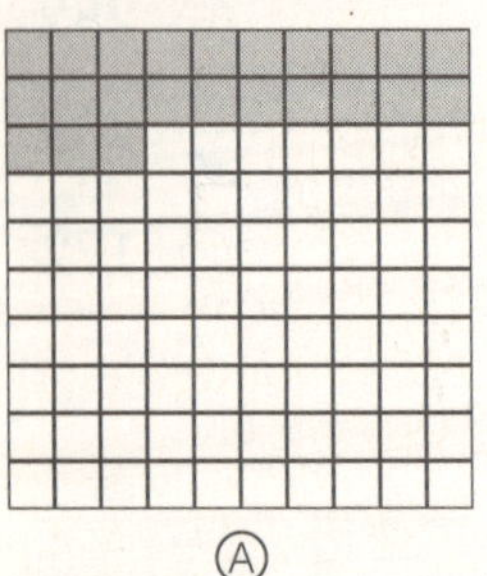

Ⓐ

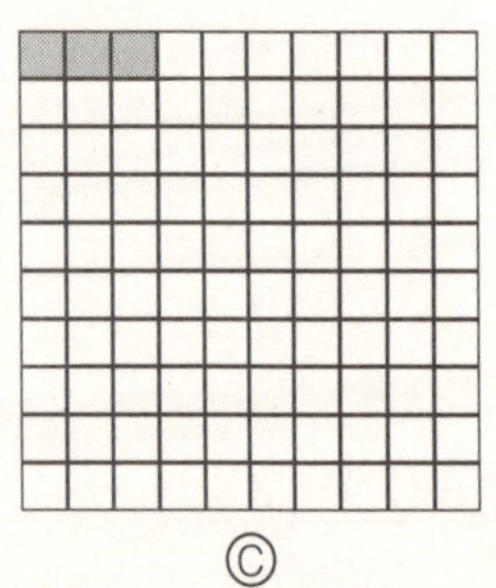

Ⓒ

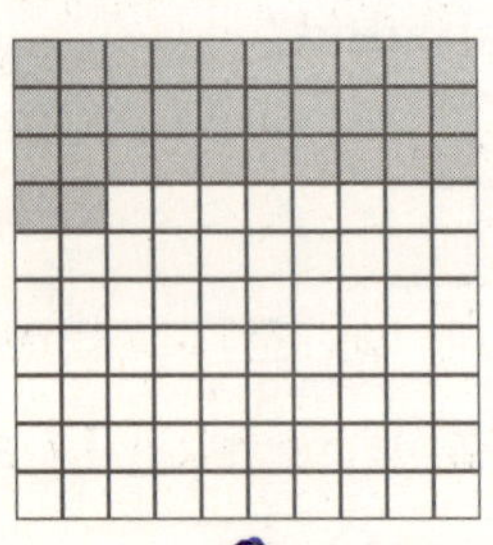

Ⓑ

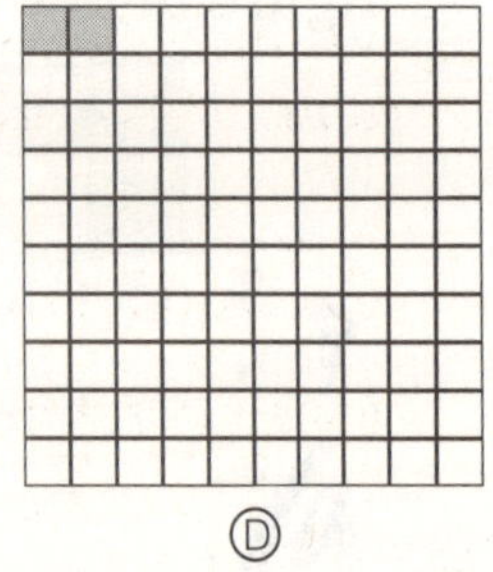

Ⓓ

❷ **Which list has the numbers in order from least value to greatest value?**

Ⓕ 8.09, 8.10, 8.23, 8.32

Ⓖ 8.10, 8.09, 8.23, 8.32

Ⓗ 8.32, 8.23, 8.10, 8.09

Ⓙ 8.09, 8.10, 8.32, 8.23

❸ **Jenna did 20 math problems in a math test. She missed 3. How many were correct? Select the correct number sentence.**

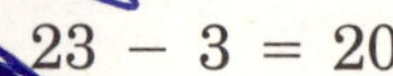

Ⓐ $20 + 3 = 23$ Ⓒ $20 - 3 = 17$

Ⓑ $23 - 3 = 20$ Ⓓ $3 \times 20 = 60$

Name ______________________________

1 Ten colored marbles were drawn from a paper bag. The bag was originally filled with 100 colored marbles. The number of marbles of each color drawn is shown in the table below

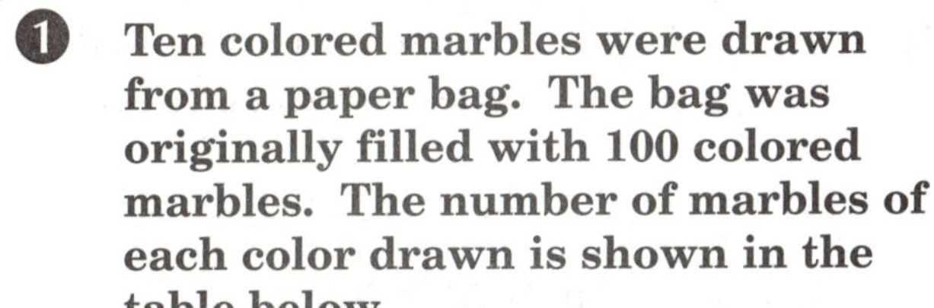

COLOR	NUMBER DRAWN
RED	6
GREEN	1
BLUE	2
YELLOW	1

If one more colored marble was drawn from the bag, which color would it most likely be?

Ⓐ red
Ⓑ green
Ⓒ blue
Ⓓ yellow

2 Look at the subtraction problems below.

$11 - 8 = \square$

$15 - 6 = \square$

$17 - 2 = \square$

If two of the numbers are switched, then the same answer will belong in all three boxes. Which two numbers should switch places?

Ⓕ switch 8 and 6
Ⓖ switch 6 and 2
Ⓗ switch 8 and 2
Ⓙ switch 11 and 2

Name ______________________________

1

15 ÷ 3 =

Ⓐ 12

Ⓑ 45

Ⓒ 3

● 5

2

IN	OUT
6	0
9	3
8	2
10	4

In the table on the left, the rule is:

Ⓕ add 5

Ⓖ subtract 6

● add 4

Ⓙ subtract 4

3

$\square \times 7 = 49$

● 6

Ⓑ 8

● 5

Ⓓ 7

4

Which number tells how many inches are in one foot?

Ⓕ 10

12

● 15

Ⓙ 16

Name ______________________________

$3 + 5 =$

Ⓐ $8 - 5$　　Ⓒ 3×5

Ⓑ $3 + 8$　　Ⓓ $5 + 3$

2 **Suppose you want to estimate the height of a telephone pole. Which one of the units of measure below would be best to use?**

Ⓕ millimeters　　Ⓗ meters

Ⓖ centimeters　　Ⓙ kilometers

3 **The line segment AB represents a:**

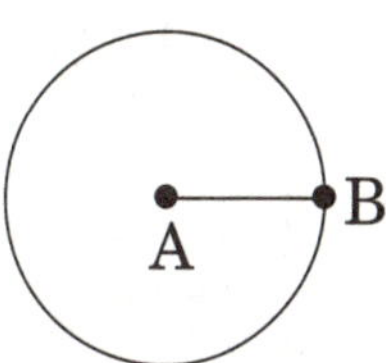

Ⓐ radius

Ⓑ diagonal

Ⓒ perimeter

Ⓓ diameter

4 $\frac{1}{4} + \frac{2}{4} =$

Ⓕ $\frac{3}{8}$　　Ⓗ $\frac{3}{4}$

Ⓖ $\frac{2}{16}$　　Ⓙ $\frac{1}{2}$

Name ______________________________

The Venn diagram below shows two sports that 17 students play.

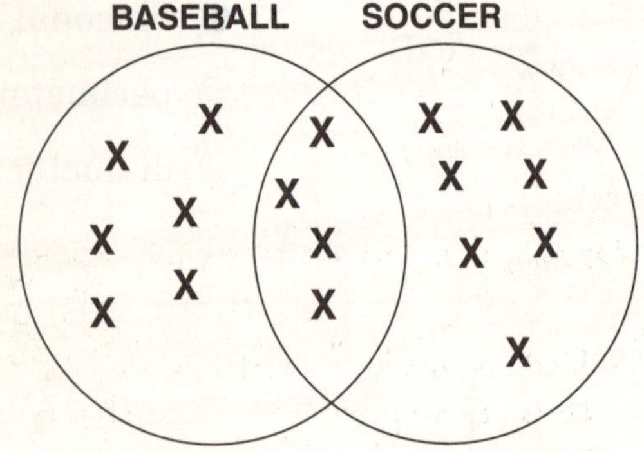

1 How many students play soccer only?

Ⓐ 6 students Ⓒ 7 students

Ⓑ 3 students Ⓓ 11 students

2 What is the total number of students that play either soccer only or both soccer and baseball?

Ⓕ 7 students Ⓗ 4 students

Ⓖ 13 students Ⓙ 11 students

3 How many students don't play soccer?

Ⓐ 4 students Ⓒ 6 students

Ⓑ 0 students Ⓓ 7 students

4 What is the total number of students that play either baseball or soccer, but not both?

Ⓕ 17 students Ⓗ 4 students

Ⓖ 13 students Ⓙ 11 students

Name ______________________________

171
× 5

 855

 Ⓑ 176

 Ⓒ 1035

Ⓓ 558

Ⓔ none of these

❸ In the figure below, is the dotted line a line of symmetry?

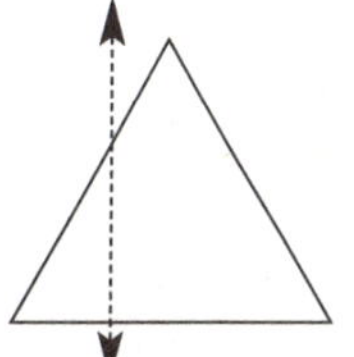

 Ⓐ yes Ⓑ no

Draw all possible lines of symmetry.

❷ Jenny borrowed a book on Wednesday. She had to return it the next day. On which day did she have to return the book?

Ⓕ Friday Ⓗ Tuesday

Ⓖ Thursday Ⓙ Saturday

❹ Which is the numeral for: 3 thousands, 4 hundreds, 5 tens, 6 ones?

Ⓕ 3456 Ⓗ 30456

Ⓖ 6543 Ⓙ 346

Name ________________________________

1 **Which of the following figures is a cone?**

Ⓐ 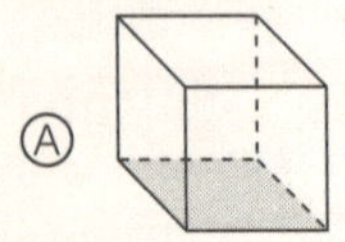

Ⓒ 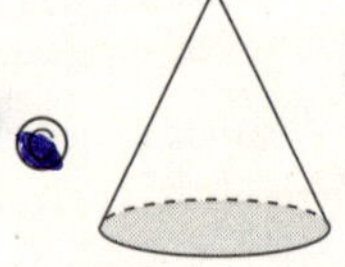

Ⓑ 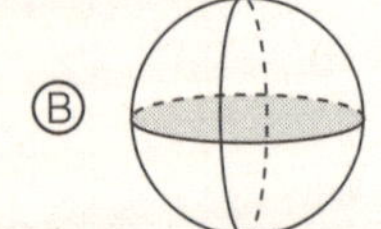

Ⓓ 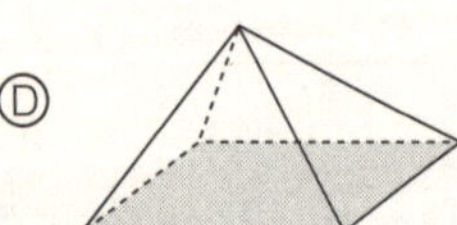

2

21 ÷ 7 =

Ⓕ 4

Ⓖ 3

Ⓗ 28

Ⓙ 148

3 **Which number names the shaded parts?**

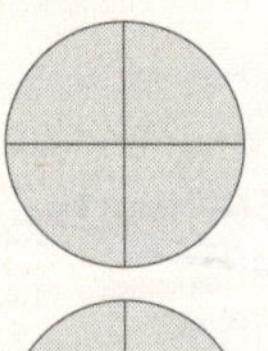

Ⓐ 7

Ⓑ $1\frac{1}{4}$

Ⓒ $1\frac{1}{8}$

Ⓓ $1\frac{3}{4}$

4 **Which number sentence goes with 8 + 4 = □?**

Ⓕ □ + 4 = 8

Ⓖ □ − 4 = 8

Ⓗ 8 ÷ 4 = □

Ⓙ □ ÷ 4 = 8

Ⓚ none of these

Name ______________________________

1 **Jessi's calendar was torn and some of the dates are missing. She is having a pool party at her house on the 2nd Sunday of the month. What is the date of her pool party?**

T	W	TH	F	S
		1	2	3
6	7	8	9	10
13	14	15	16	17
20	21	22	23	24
27	28	29	30	31

Ⓐ 4
Ⓑ 10
Ⓒ 11
Ⓓ 17

2

$$7 \times \square = 56$$

Ⓕ 9
Ⓖ 7
Ⓗ 8
Ⓙ 49

3

$$30\overline{)630}$$

Ⓐ 201
Ⓑ 21
Ⓒ 20
Ⓓ 20 R1

Name ______________________________

1

$$\begin{array}{r} 503 \\ \times \quad 2 \\ \hline \end{array}$$

Ⓐ 1006

Ⓑ 505

Ⓒ 1026

Ⓓ 1016

2

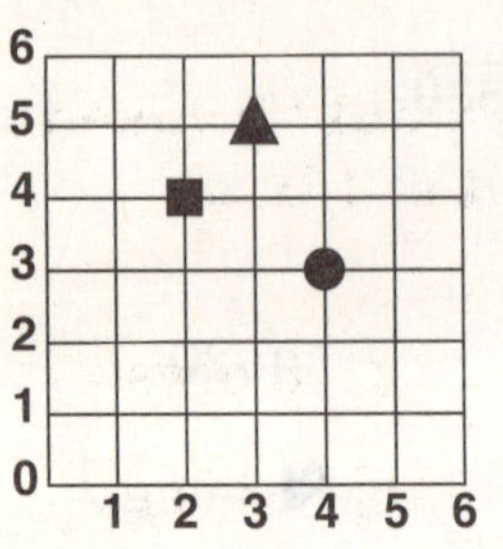

Look at the graph. What is located at (3, 5)?

Ⓕ ●

Ⓖ ▲

Ⓗ ■

3 **There were 210 children at school. 55 children left school to go to the park. How many children remained at school? Select the correct number sentence.**

Ⓐ 210 + 55 = n

Ⓑ 210 − 55 = n

Ⓒ 55 + 210 = n

Ⓓ 210 ÷ 55 = n

4

$$\frac{3}{6} + \frac{2}{6} =$$

Ⓕ $\frac{5}{6}$

Ⓖ $\frac{5}{12}$

Ⓗ $\frac{1}{6}$

Ⓙ $\frac{6}{36}$

Name ______________________________

1 **5020 + 2165 + 1214 =**

Ⓐ 8389

Ⓑ 8299

Ⓒ 8399

Ⓓ 8289

Ⓔ none of these

2 **(9 × 6) × 3 is the same as:**

Ⓕ 9 × (6 × 3)

Ⓖ (9 × 9) × 3

Ⓗ (9 × 18) × 6

Ⓙ (9 × 6) × 27

3 **The figure below shows that the 2 boxes weigh the same as 6 eggs.**

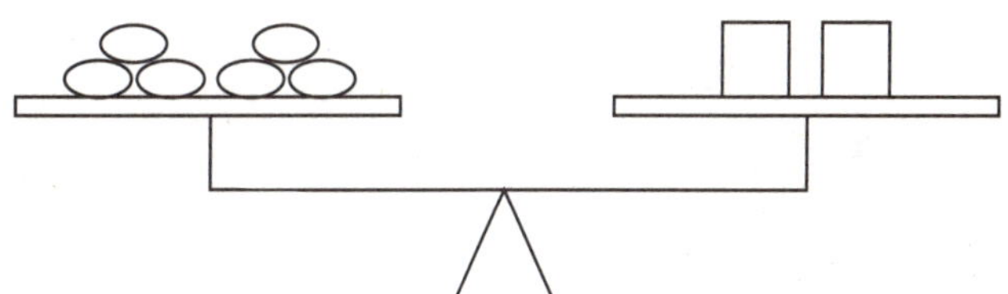

How many eggs would it take to equal the weight of 1 box?

Ⓐ 2 eggs

Ⓑ 3 eggs

Ⓒ 6 eggs

Ⓓ 12 eggs

Name ______________________________

1 **The graph shows the temperature at 2:00 P.M. on July 4th for five years in a row. Which one of the following is <u>not</u> a true statement about the temperature from 1991 to 1995?**

Ⓐ decreased between 1992 and 1994

Ⓑ the same temperature in 1991 and 1993

Ⓒ the greatest change in one year occurred between 1992 and 1993

Ⓓ the coldest in 1994

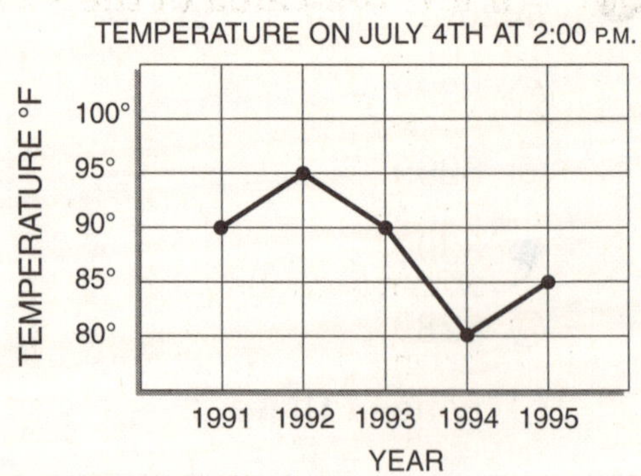

---

2 **4 dimes + 1 nickel + 2 pennies =**

Ⓕ 57¢

Ⓖ 47¢

Ⓗ 67¢

Ⓙ 87¢

3

$3\overline{)17}$

Ⓐ 5 R2

Ⓑ 4 R2

Ⓒ 5 R1

Ⓓ 8 R1

Name ______________________________

1 **What is the value of the 8 in 30.86?**

Ⓐ 8 tens

Ⓑ 8 ones

Ⓒ 8 hundredths

Ⓓ 8 tenths

2

$$\begin{array}{r} \mathbf{365} \\ \times \quad \mathbf{4} \\ \hline \end{array}$$

Ⓕ 1440

Ⓖ 1622

Ⓗ 1620

Ⓙ 1460

3

$$\frac{3}{4} - \frac{2}{4} =$$

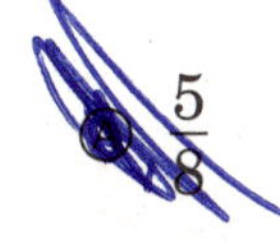

Ⓐ $\frac{5}{8}$

Ⓒ $\frac{1}{0}$

Ⓑ $\frac{1}{4}$

Ⓓ $\frac{6}{16}$

4 **What time is shown?**

Ⓕ 4:50

Ⓖ 4:20

Ⓗ 5:10

Ⓙ 4:40

Name ______________________

1

$$\begin{array}{r} 501 \\ -\ 188 \\ \hline \end{array}$$

Ⓐ 487

Ⓑ 313

Ⓒ 407

Ⓓ 423

2

IN	OUT
2	4
3	6
4	
5	10

Look at the table. The missing number is:

Ⓕ 7

Ⓖ 8

Ⓗ 9

Ⓙ 5

3

$4\overline{)29}$

Ⓐ 8 R5

Ⓑ 7 R1

Ⓒ 7 R2

Ⓓ 9 R2

4 **What is the value of the 3 in 64.31?**

Ⓕ 3 tenths

Ⓖ 3 hundredths

Ⓗ 3 ones

Ⓙ 3 tens

Name ______________________________

1 **How many inches are in 3 feet?**

Ⓐ 30 inches
Ⓑ 36 inches
Ⓒ 48 inches
Ⓓ 15 inches

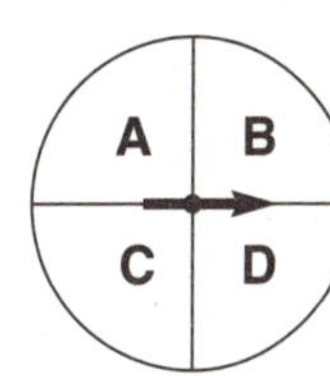

On one spin of the spinner, the probability it will stop on B is:

Ⓐ $\frac{3}{4}$
Ⓑ 0
Ⓒ $\frac{1}{4}$
Ⓓ $\frac{1}{2}$

$$\begin{array}{r} \$\,44.30 \\ -\ 13.29 \\ \hline \end{array}$$

Ⓕ \$57.51
Ⓖ \$31.01
Ⓗ \$31.11
Ⓙ \$31.19
Ⓚ none of these

4 $(3 \times 1000) + (6 \times 100) + (7 \times 10) + (3 \times 1) =$

Ⓕ 3673
Ⓖ 30,673
Ⓗ 367
Ⓙ 373

Name ______________________________

1

$$6\overline{)44}$$

Ⓐ 6 R2
Ⓑ 7 R2
Ⓒ 8 R2
Ⓓ 7 R4

2 **Which number sentence goes with 11 − 6 = □?**

Ⓕ □ − 6 = 11
Ⓖ □ + 6 = 11
Ⓗ 11 + 6 = □
Ⓙ □ ÷ 6 = 11

3 **A basket of strawberries costs $4. If a strawberry stand sold 84 baskets in one weekend, how much did they earn?**

Ⓐ $320
Ⓑ $326
Ⓒ $336
Ⓓ $346

4

$$\frac{3}{6} - \frac{2}{6} =$$

Ⓕ $\frac{1}{6}$
Ⓖ $\frac{5}{12}$
Ⓗ $\frac{1}{0}$
Ⓙ $\frac{6}{36}$

Name ______________________________

1. **There are 15 grapes in a bowl. Three girls eat all the grapes. Each girl eats at least 1, but fewer than 8. List all the combinations the three girls could have eaten.**

2. **In the problem above, would it be possible for each girl to eat an <u>even</u>, not odd, number of grapes? Explain why or why not.**

Name ______________________________

1 **Which number sentence goes with $6 \times 2 = \square$?**

- (A) $\square \times 2 = 6$
- (B) $\square + 2 = 6$
- (C) $6 \div 2 = \square$
- (D) $\square \div 2 = 6$

2 **6010 is read as:**

- (F) six thousand one
- (G) sixty thousand ten
- (H) sixty ten
- (J) six thousand ten

3 **What is the missing number?**

200, ______, 400, 500

- (A) 250
- (B) 300
- (C) 600
- (D) 100

4

$7\overline{)86}$

- (F) 11 R1
- (G) 12 R2
- (H) 12 R4
- (J) 13

Name ______________________________

1 **David's family loves to eat eggs prepared many different ways. The pictograph shows the number of eggs that David's family ate during one month. How many eggs did they eat in week 3?**

Ⓐ 12 eggs
Ⓑ 18 eggs
Ⓒ 24 eggs
Ⓓ 15 eggs

EGGS EATEN BY DAVID'S FAMILY

WEEK	NUMBER EATEN
1	▦ ▦
2	▦ ▦ ◧
3	▦ ◧
4	▦

▦ = 12 EGGS

2 **The factors (divisors) of 4 are:**

Ⓕ 1, 2, 4
Ⓖ 0, 1, 4
Ⓗ 0, 1, 2, 4
Ⓙ 4, 8, 12

3 **How many feet are in one yard?**

Ⓐ 3 feet
Ⓑ 4 feet
Ⓒ 2 feet
Ⓓ 36 feet

Name ______________________________

$$\begin{array}{r} 324 \\ \times \quad 6 \\ \hline \end{array}$$

Ⓐ 1824

Ⓑ 1026

Ⓒ 1904

Ⓓ 1964

Ⓔ none of these

2 **There are 12 stickers. If they share them equally, how many each may 4 children have?**

Ⓕ 48 stickers

Ⓖ 4 stickers

Ⓗ 8 stickers

Ⓙ 3 stickers

3 **Round 7991 to the nearest thousand.**

Ⓐ 8000

Ⓑ 7000

Ⓒ 9000

Ⓓ 7500

4 $\frac{7}{8} - \frac{6}{8} =$

Ⓕ $\frac{1}{8}$

Ⓖ $\frac{13}{16}$

Ⓗ $\frac{1}{0}$

Ⓙ $\frac{42}{64}$

Name ______________________________

$$\frac{5}{9} + \frac{2}{9} =$$

Ⓐ $\frac{7}{9}$

Ⓑ $\frac{7}{18}$

Ⓒ $\frac{2}{3}$

Ⓓ $\frac{1}{3}$

2 **Lauren has collected 60 antique dolls. She wants to put an equal number of dolls into 5 storage boxes. How many dolls should she put into each box?**

Ⓕ 300 dolls

Ⓖ 30 dolls

Ⓗ 10 dolls

Ⓙ 12 dolls

3 **Which of the drawings below is a right angle?**

Ⓐ 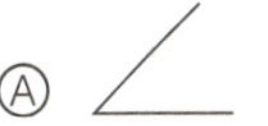

Ⓒ 

Ⓑ 

Ⓓ 

4 **Look at the graph. Where is the ● ?**

Ⓕ (2, 5)

Ⓖ (1, 2)

Ⓗ (3, 6)

Name ________________________________

1 **Suppose you want to estimate the weight of a horse. Which one of the units of measure below would be best to use?**

Ⓐ ounces
Ⓑ pounds
Ⓒ tons
Ⓓ grams

2 **The factors (divisors) of 8 are:**

Ⓕ 0, 1, 8
Ⓖ 1, 2, 4, 8
Ⓗ 2, 4, 6
Ⓙ 1, 2, 8, 12

3 **What shape would come next in the pattern?**

Ⓐ ★
Ⓑ ✚
Ⓒ ▲

4 **What is $79.85 rounded to the nearest dollar?**

Ⓕ $79.00
Ⓖ $70.00
Ⓗ $100.00
Ⓙ $80.00

Name ______________________________

1 **Jenny, Christy, and Scott weighed themselves. Their weights are 65, 72, and 80 pounds. Christy's weight in <u>not</u> 72 pounds. Scott's weight is <u>not</u> 80 pounds. Christy weighs <u>less</u> than Scott. How much does Jenny weigh? You may want to use the table below to help you find the answer.**

	65	72	80
JENNY			
CHRISTY			
SCOTT			

Jenny weighs ________ lbs.

Name ______________________________

1

$$\begin{array}{r} \$\,56.48 \\ -\ 18.29 \\ \hline \end{array}$$

Ⓐ $38.19

Ⓑ $38.29

Ⓒ $42.21

Ⓓ $74.77

2 **Which number is an even number?**

Ⓕ 101

Ⓖ 103

Ⓗ 102

Ⓙ 105

3 **What is the value of the 5 in 23.05?**

Ⓐ 5 ones

Ⓑ 5 tenths

Ⓒ 5 hundredths

Ⓓ 5 thousandths

4

IN	OUT
6	13
9	16
12	19
19	26

In the table on the left, the rule is:

Ⓕ add 6

Ⓖ subtract 7

Ⓗ add 7

Ⓙ none of these

Name ______________________________

1

$$\begin{array}{r} 21 \\ \times\ 11 \\ \hline \end{array}$$

Ⓐ 221

Ⓑ 231

Ⓒ 42

Ⓓ 22 R1

Ⓔ none of these

2 **Which of the figures below is <u>not</u> a polygon?**

Ⓕ 

Ⓗ 

Ⓖ 

Ⓙ

3 **Alyssa and her grandfather were born on the 11th day of October. If Alyssa's grandfather was born in 1925 and Alyssa was 3 years old in 1990, how old was Alyssa's grandfather when she was born?**

Ⓐ 65 years old

Ⓒ 63 years old

Ⓑ 62 years old

Ⓓ 68 years old

4 $\frac{8}{9} - \frac{3}{9} =$

Ⓕ $\frac{4}{9}$

Ⓗ $\frac{11}{18}$

Ⓖ $\frac{5}{0}$

Ⓙ $\frac{5}{9}$

Name ______________________________

1

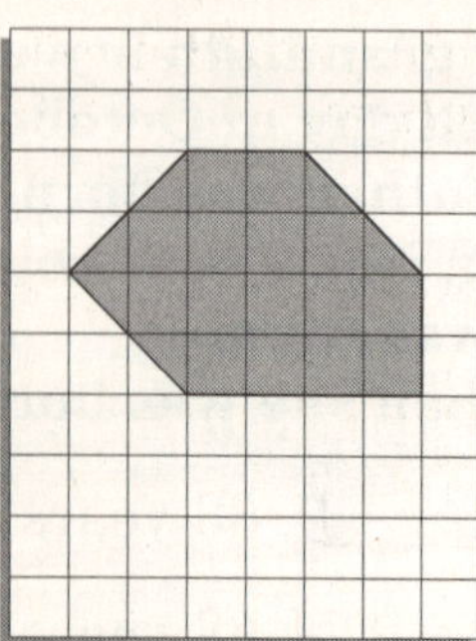

The shaded figure on the grid to the left has an area of about:

Ⓐ 10 square units.

Ⓑ 13 square units.

Ⓒ 15 square units.

Ⓓ 18 square units.

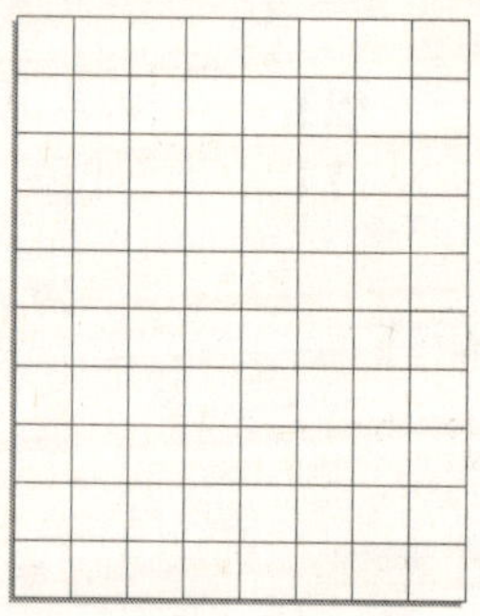

Draw a rectangle on the blank grid to the right that has the same area.

2

$$\begin{array}{r} \$\ 1.31 \\ \times\ \ \ \ 2 \\ \hline \end{array}$$

Ⓕ $2.62

Ⓖ $2.72

Ⓗ $1.33

Ⓙ $20.62

3 **The factors (divisors) of 12 are:**

Ⓐ 1, 2, 5, 6

Ⓑ 0, 2, 4, 6

Ⓒ 1, 2, 3, 4, 6, 12

Ⓓ 0, 1, 4, 7, 9

Name ______________________________

$7 \times \square = 42$

Ⓐ 6 Ⓒ 294

Ⓑ 7 Ⓓ 35

Three multiples of 2 are:

Ⓕ 1, 2, 3 Ⓗ 1, 2, 5

Ⓖ 1, 2, 4 Ⓙ 2, 4, 6

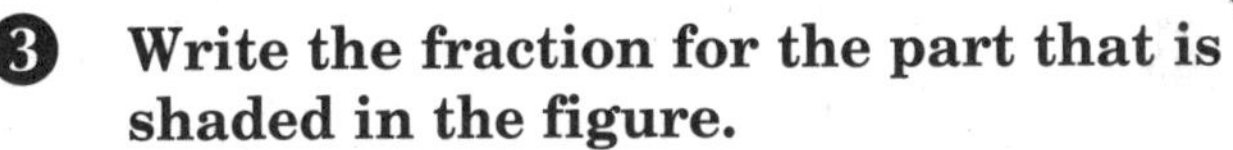

Write the fraction for the part that is shaded in the figure.

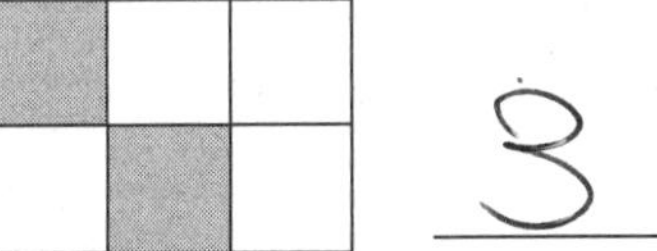

❹ Write an equivalent fraction to your answer in question 3. ________

Shade the figure to match the equivalent fraction.

Name ______________________________

$$\frac{4}{6} + \frac{1}{6} =$$

Ⓐ $\frac{5}{12}$ Ⓒ $\frac{5}{6}$

Ⓑ $\frac{1}{2}$ Ⓓ $\frac{5}{10}$

2 **What is the average of these numbers?**

(1, 2, 5, 8)

Ⓕ 4 Ⓗ 9

Ⓖ 16 Ⓙ 6

3 **On one spin of the spinner, the probability it will stop on 3 is:**

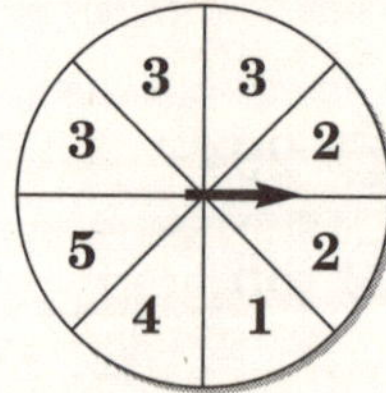

Ⓐ $\frac{5}{8}$ Ⓒ $\frac{1}{4}$

Ⓑ $\frac{3}{8}$ Ⓓ $\frac{2}{8}$

4

$$\begin{array}{r} \$1.43 \\ \times \quad 2 \\ \hline \end{array}$$

Ⓕ $2.96

Ⓖ $1.45

Ⓗ $2.86

Ⓙ $3.86

Name ________________________________

1 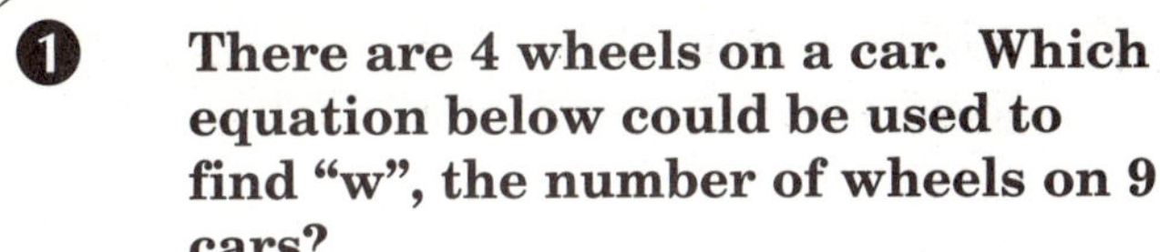

There are 4 wheels on a car. Which equation below could be used to find "w", the number of wheels on 9 cars?

Ⓐ $9 + 4 = w$

Ⓑ $4 \times w = 9$

Ⓒ $9 \div 4 = w$

Ⓓ $9 \times 4 = w$

Ⓔ none of these

2

$$\square \times \mathbf{8} = \mathbf{24}$$

Ⓕ 4

Ⓖ 32

Ⓗ 3

Ⓙ 192

3 **Tinh walked 3 blocks to the drug store and spent \$6.49 for film. Later, he walked 5 blocks to the dairy and spent \$7.15 for milk. How much money did Tinh spend?**

Ⓐ \$21.64

Ⓑ \$8.00

Ⓒ \$12.15

Ⓓ \$13.64

4

$$3\overline{)16}$$

Ⓕ 5 R1

Ⓖ 5

Ⓗ 4

Ⓙ 53 R1

Name ____________________

1 **Which grid has forty-six hundredths shaded?**

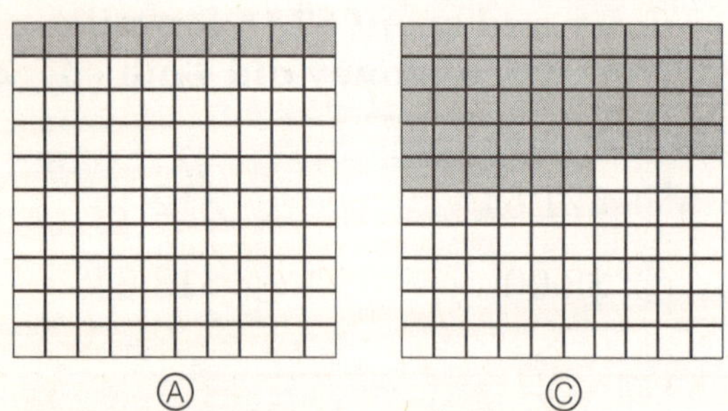

Ⓐ Ⓒ

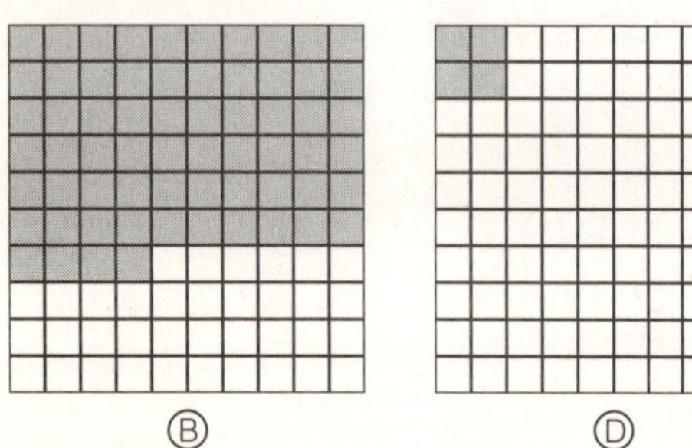

Ⓑ Ⓓ

2 **What is the value of the 8 in 10.86?**

Ⓕ 8 ones

Ⓖ 8 tens

Ⓗ 8 hundredths

Ⓙ 8 tenths

3 **Cori, Dani, Maria, and Jill each ran one lap around the school track. Cori was 10 seconds faster than Dani. Maria was 5 seconds slower than Jill. Dani was 5 seconds faster than Jill. Which one of the following statements is true about how their times compared?**

Ⓐ Dani was the fastest.

Ⓑ Dani and Jill had the same time.

Ⓒ Jill was third fastest.

Ⓓ Cori was second fastest.

Name ______________________________

 **Distance from San Diego, CA to:**

New York, NY	2821 miles
Chicago, IL	2056 miles
Atlanta, GA	2132 miles
Washington, D.C.	2668 miles

Which list shows the cities in order from nearest to farthest from San Diego, CA?

Ⓐ New York, Washington, Atlanta, Chicago

Ⓑ Chicago, Atlanta, Washington, New York

Ⓒ Chicago, Atlanta, New York, Washington

Ⓓ Atlanta, Chicago, Washington, New York

❷ **What is the missing number?**

9, 18, ____, 36

Ⓕ 35

Ⓖ 27

Ⓗ 19

Ⓙ 26

❸ **Which number is divisible by 9?**

Ⓐ 34

Ⓑ 35

Ⓒ 36

Ⓓ 37

Name ______________________________

1 **Robert, James, and Luis have a total of 18 baseball cards between them. Each boy has at least one card. Each has a different number of cards. Robert has one more card than James. Luis has the most cards.**

Find as many different combinations as you can to show how many cards each boy could have. Show your work.

Name ______________________________

1

$$\begin{array}{r} 21 \\ \times\ 13 \\ \hline \end{array}$$

Ⓐ 84

Ⓑ 273

Ⓒ 34

Ⓓ 651

2 **7 + 8 = 15**

8 + 7 = 15

15 − 7 = 8

Which goes with the three above?

Ⓕ 15 − 15 = 0

Ⓖ 15 − 8 = 7

Ⓗ 8 + 15 = 23

Ⓙ 15 − 0 = 15

3 **Five friends ordered hamburgers at the ball game. Four hamburgers had catsup. Only two had both catsup and mustard. How many hamburgers had mustard only? You may use the Venn diagram below or any other drawing to find the answer.**

Ⓐ 0 hamburgers

Ⓑ 1 hamburger

Ⓒ 2 hamburgers

Ⓓ 3 hamburgers

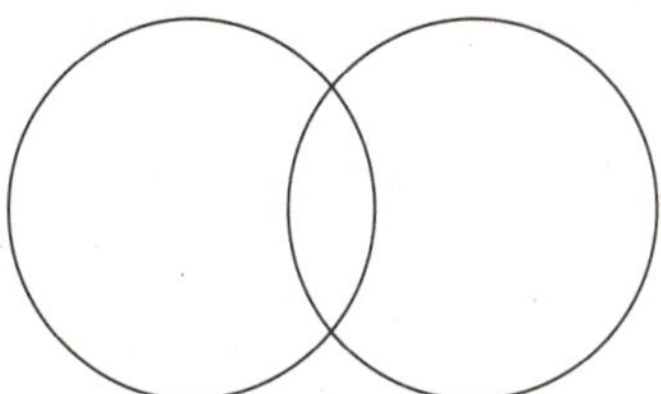

Name ______________________________

1

$$\begin{array}{r} \frac{1}{5} \\ + \frac{1}{5} \\ \hline \end{array}$$

Ⓐ $\frac{2}{5}$ Ⓒ $\frac{2}{10}$

Ⓑ $\frac{1}{25}$ Ⓓ $\frac{1}{5}$

3 **In the figure below, is the dotted line a line of symmetry?**

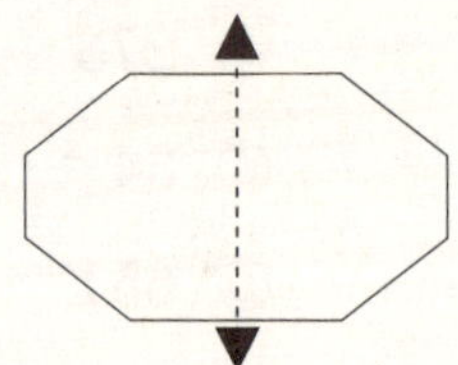

Ⓐ yes Ⓑ no

Draw all possible lines of symmetry.

2 **You arrived at the airport at 10:15 P.M. but found out that your plane had left at 9:35 P.M. By how many minutes did you miss your plane?**

Ⓕ 30 minutes Ⓗ 20 minutes

Ⓖ 40 minutes Ⓙ 50 minutes

4

$$\begin{array}{r} \$2.15 \\ \times \quad 3 \\ \hline \end{array}$$

Ⓕ $6.45

Ⓖ $7.45

Ⓗ $6.35

Ⓙ $6.48

Ⓚ none of these

Name ______________________________

1 $(1 \times 1000) + (5 \times 100) + (8 \times 10) + (1 \times 1) =$

Ⓐ 10,581
Ⓑ 1,581
Ⓒ 1,851
Ⓓ 1,580

2

$$4\overline{)23}$$

Ⓕ 5 R4
Ⓖ 4 R3
Ⓗ 5 R3
Ⓙ 4 R4

3

$$\begin{array}{r} 21 \\ \times\ 14 \\ \hline \end{array}$$

Ⓐ 105
Ⓑ 35
Ⓒ 284
Ⓓ 294

4 What is the area of the figure below?

5 cm

5 cm

Ⓕ 5 sq. cm
Ⓖ 10 sq. cm
Ⓗ 20 sq. cm
Ⓙ 25 sq. cm

Name ______________________________

1 One piece of ribbon is 9 cm long. Another piece of ribbon is 12 cm long. If the two pieces of ribbon are taped together to make a piece 17 cm long, how long is the overlap?

Ⓐ 8 cm Ⓑ 4 cm Ⓒ 3 cm Ⓓ 5 cm

Draw a picture to prove your answer.

2 Which of the drawings below shows intersecting lines?

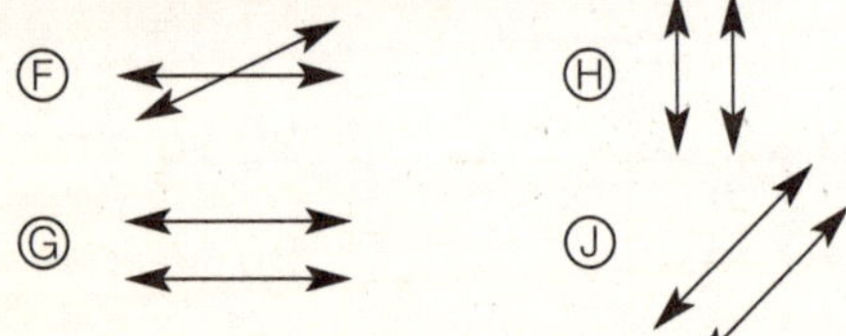

3 Look at the graph. What is located at (C, 3)?

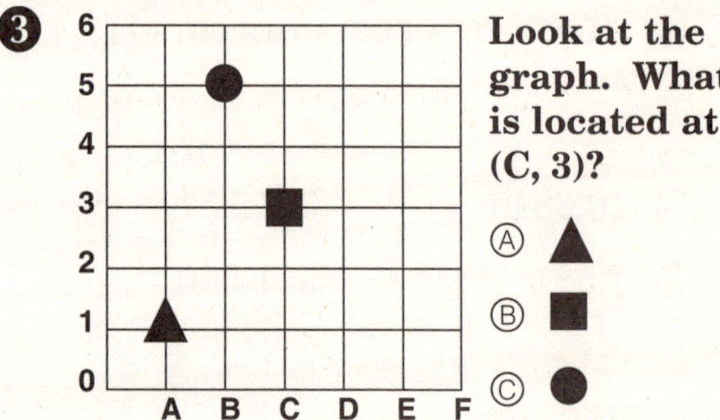

Ⓐ ▲

Ⓑ ■

Ⓒ ●

4 There are 10 children in line. Maria is third. How many are <u>behind</u> her?

Ⓕ 10 Ⓗ 7

Ⓖ 9 Ⓙ 2

Name ______________________________

1

$$\begin{array}{r} 25 \\ \times\ 10 \\ \hline \end{array}$$

Ⓐ 205

Ⓑ 250

Ⓒ 25

Ⓓ 2050

2 **7004 is read as:**

Ⓕ seven hundred four

Ⓖ seventy-four hundred

Ⓗ seventy-four thousand

Ⓙ seven thousand four

3 **Using each of the numerals 3, 6, 7, and 8 one time only, what is the largest number you can make:**

with 8 in the tens place? __________

with 8 in the hundreds place? ______

with 8 in the thousands place? ______

Using the numbers you wrote, find the difference between the number of greatest value and the number of least value.

____

Name ______________________________

$3\overline{)17}$

Ⓐ 5 R2

Ⓑ 6

Ⓒ 5 R1

Ⓓ 50 R2

2 **Round 51 to the nearest hundred.**

Ⓕ 0

Ⓖ 50

Ⓗ 100

Ⓙ 500

$$\begin{array}{r} \$\ 2.34 \\ \times \quad 3 \\ \hline \end{array}$$

Ⓐ $6.02

Ⓑ $7.92

Ⓒ $6.92

Ⓓ $7.02

4 **By drawing only 3 straight lines on the hexagon to the left, you can make it become a cube! Can you do it?**

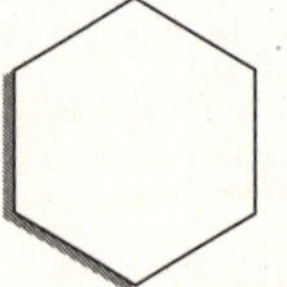

Hexagon

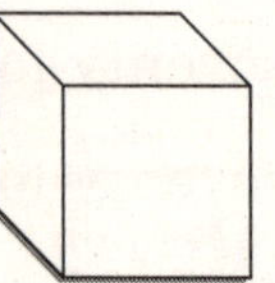

Cube

Name ________________________________

1

A

B

Which of the two spinners gives you the best chance of spinning the number 1?

Ⓐ A　　Ⓑ B

Explain your reasoning:

2

$$\begin{array}{r} 23 \\ \times\ 23 \\ \hline \end{array}$$

Ⓕ 4669

Ⓖ 46

Ⓗ 1006

Ⓙ 539

Ⓚ none of these

3

$$\begin{array}{r} 605 \\ -\ 515 \\ \hline \end{array}$$

Ⓐ 115

Ⓑ 110

Ⓒ 90

Ⓓ 1120

Ⓔ none of these

Name ______________________________

1

$$\begin{array}{r} \$\ 55.00 \\ -\ 23.78 \\ \hline \end{array}$$

Ⓐ \$31.21

Ⓑ \$32.78

Ⓒ \$31.22

Ⓓ \$32.22

2 **What is the greatest fraction?**

Ⓕ $\frac{3}{12}$

Ⓗ $\frac{1}{3}$

Ⓖ $\frac{1}{5}$

Ⓙ $\frac{1}{2}$

3

$$\frac{8}{10} - \frac{1}{10} =$$

Ⓐ $\frac{9}{10}$

Ⓒ $\frac{7}{10}$

Ⓑ $\frac{7}{0}$

Ⓓ $\frac{4}{5}$

4 **Fifty-one students and four adults are going to take a bus to the zoo. The bus has twenty-eight seats and each seat can hold two people. Are there enough seats?**

____

Will any of the seats have fewer than two people? If so how many?

____

Name ____________________

$$\begin{array}{r} \$2.13 \\ \times \quad 3 \\ \hline \end{array}$$

Ⓐ $6.49

Ⓑ $2.16

Ⓒ $6.39

Ⓓ $6.42

IN	OUT
3	21
5	35
0	0
	42

Look at the table. The missing number is:

Ⓕ 4

Ⓖ 7

Ⓗ 10

Ⓙ 6

❸ **A brick wall is being built. There are 40 bricks in the first row, 34 bricks in the second row, and 28 bricks in the third row. If this pattern continues, how many bricks will be in the <u>fifth</u> row?**

Ⓐ 102 bricks

Ⓑ 26 bricks

Ⓒ 22 bricks

Ⓓ 16 bricks

Explain your answer.

Name ____________________

1 **Find the factors (divisors) of 9.**

Ⓐ 2, 3, 6

Ⓑ 1, 3, 9

Ⓒ 9, 18, 27

Ⓓ 0, 3, 9

2 **There are 28 students in Roberto's class. One day 3 boys and 2 girls were absent. How many children were not absent?**

Ⓕ 25 children

Ⓖ 26 children

Ⓗ 33 children

Ⓙ 23 children

3

$$8 \times \square = 72$$

Ⓐ 8

Ⓑ 9

Ⓒ 576

Ⓓ 80

4

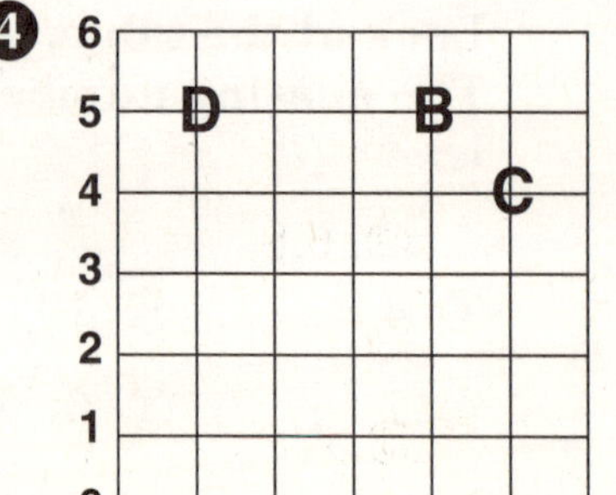

Which letter names the point (4, 5)?

Ⓕ C

Ⓖ B

Ⓗ D

Name ______________________________

1

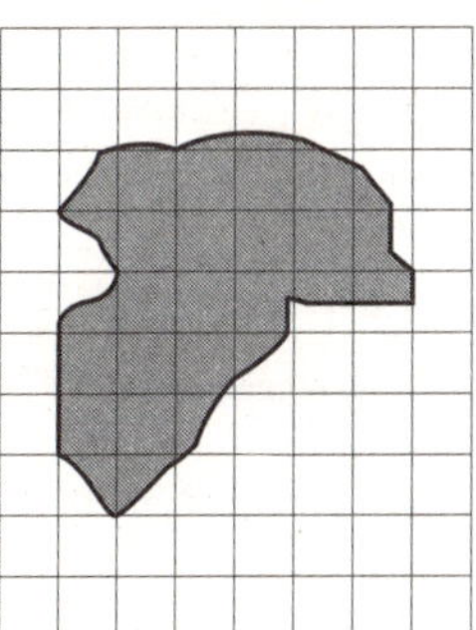

The shaded figure on the grid to the left has an area of about:

Ⓐ 15 square units.
Ⓑ 19 square units.
Ⓒ 22 square units.
Ⓓ 25 square units.
Ⓔ none of these

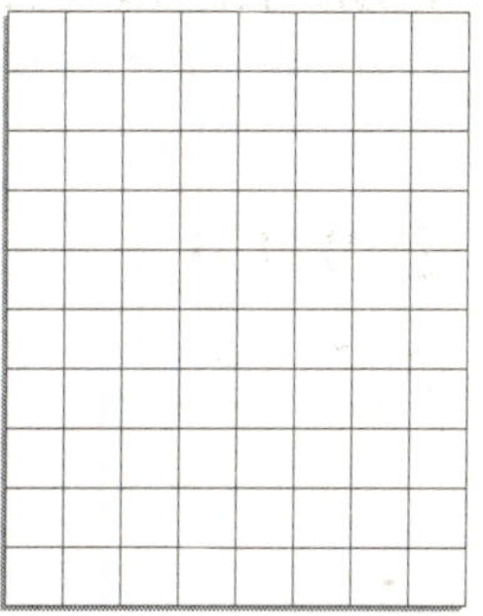

On the blank grid to the right, draw a rectangle that is one unit <u>less</u> in area.

2

$$\begin{array}{r} 3.4 \\ +\ 2.2 \\ \hline \end{array}$$

Ⓕ 5.6
Ⓖ 6.8
Ⓗ 1.2
Ⓙ 5.5
Ⓚ none of these

3

$$\begin{array}{r} 21 \\ \times\ 24 \\ \hline \end{array}$$

Ⓐ 484
Ⓑ 126
Ⓒ 504
Ⓓ 514
Ⓔ none of these

Name ______________________________

The data in the table below shows the number of plants growing in Mr. Green's garden.

PLANTS	NUMBER GROWING IN GARDEN
CARROTS	10
LETTUCE	4
TOMATOES	8
BEANS	12

❶ **Draw a graph to show this data.**

❷ **Why is a graph a useful way to record and show data?**

Name ______________________________

1

$4\overline{)845}$

Ⓐ 201 R1 Ⓒ 211 R4

Ⓑ 211 R1 Ⓓ 210 R1

2

$$\begin{array}{r} 5.6 \\ +\ 1.3 \\ \hline \end{array}$$

Ⓕ 4.3

Ⓖ 7.9

Ⓗ 6.8

Ⓙ 6.9

3 **You are planning to meet some friends at the park at 3:00 P.M. It will take you exactly 25 minutes to walk there. If it is 2:00 P.M., how much time is left before you must leave to arrive at the park at 3:00 P.M.?**

Ⓐ 1 hour Ⓒ 35 minutes

Ⓑ 25 minutes Ⓓ 45 minutes

4

$$\begin{array}{r} \$2.15 \\ \times\quad 3 \\ \hline \end{array}$$

Ⓕ $6.45

Ⓖ $7.45

Ⓗ $6.35

Ⓙ $2.18

Name ______________________________

1 **Three multiples of 9 are:**

Ⓐ 9, 18, 27　Ⓒ 3, 6, 9

Ⓑ 1, 3, 9　Ⓓ 3, 9,12

2 **Talat has 3 quarters, 2 dimes and 1 nickel and wants to buy an ice-cream cone for 68¢. What combination of coins should Talat use to buy the ice-cream cone and receive the least amount of change?**

Ⓕ 3 quarters

Ⓖ 2 quarters, 1 dime, and 1 nickel

Ⓗ 2 quarters, 2 dimes, and 1 nickel

Ⓙ 2 quarters and 2 dimes

3

$$\begin{array}{r} 2251 \\ \times \quad 3 \\ \hline \end{array}$$

Ⓐ 7113

Ⓑ 6753

Ⓒ 6654

Ⓓ 7753

4 **Which number is divisible by 2 <u>and</u> 3?**

Ⓕ 5　Ⓗ 8

Ⓖ 6　Ⓙ 9

Name ______________________________

1 **5 classes shared 15 gallons of punch. How many gallons of punch did each class get?**

Ⓐ 75 gallons
Ⓑ 3 gallons
Ⓒ 5 gallons
Ⓓ 20 gallons

2

$$\begin{array}{r} 23 \\ \times\ 22 \\ \hline \end{array}$$

Ⓕ 45
Ⓖ 506
Ⓗ 92
Ⓙ 496
Ⓚ none of these

3

$$\begin{array}{r} 7.1 \\ +\ 1.5 \\ \hline \end{array}$$

Ⓐ 8.06
Ⓑ 5.6
Ⓒ 9.6
Ⓓ 6.4
Ⓔ none of these

---

4 **In the figure below, is the dotted line a line of symmetry?**

Ⓕ yes
Ⓖ no

Draw all possible lines of symmetry.

Name ____________________

1 **Which list has the numbers in order from least value to greatest value?**

Ⓐ 8.09, 8.10, 8.23, 8.32

Ⓑ 8.10, 8.09, 8.23, 8.32

Ⓒ 8.32, 8.23, 8.10, 8.09

Ⓓ 8.09, 8.10, 8.32, 8.23

2 **How many inches in $\frac{1}{2}$ foot?**

Ⓕ 12 inches

Ⓖ 6 inches

Ⓗ 5 inches

Ⓙ 8 inches

3 **The figure below shows that the 5 oranges weigh the same as 15 blocks.**

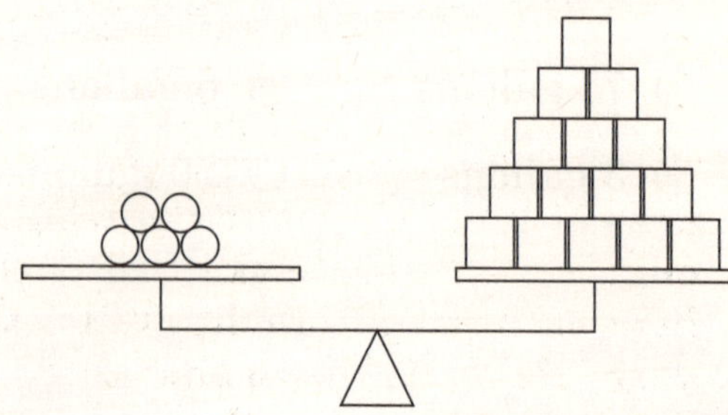

How many blocks would it take to equal the weight of 2 oranges?

Ⓐ 3 blocks

Ⓑ 6 blocks

Ⓒ 15 blocks

Ⓓ 50 blocks

Name ______________________________

1

$$\begin{array}{r} 8.6 \\ -\ 3.4 \\ \hline \end{array}$$

Ⓐ 12.0

Ⓑ 4.2

Ⓒ 5.02

Ⓓ 5.2

2 **Jose, Steve, Peggy and Sal want to choose teams for basketball based on height. Steve is 2 inches taller than Jose. Peggy is 1 inch shorter than Sal but 1 inch taller than Jose. Which of the following statements is <u>not</u> true?**

Ⓕ Peggy is shorter than Steve.

Ⓖ Steve is taller than Jose, Peggy and Sal.

Ⓗ Jose is the shortest.

Ⓙ Sal and Steve are the same height.

3

$3\overline{)609}$

Ⓐ 213

Ⓑ 206 R1

Ⓒ 203

Ⓓ 302

4

$$\begin{array}{r} \$2.34 \\ \times \quad 3 \\ \hline \end{array}$$

Ⓕ $69.12

Ⓖ $6.92

Ⓗ $7.12

Ⓙ $7.02

Name ______________________________

$$\frac{3}{7} + \frac{1}{7}$$

Ⓐ $\frac{4}{7}$　　Ⓒ $\frac{3}{0}$

Ⓑ $\frac{4}{14}$　　Ⓓ $\frac{2}{7}$

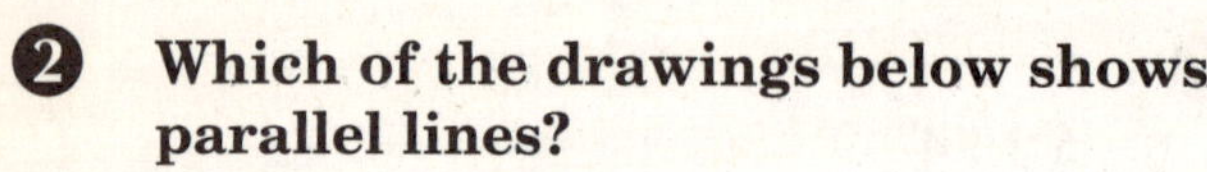

❷ **Which of the drawings below shows parallel lines?**

Ⓕ 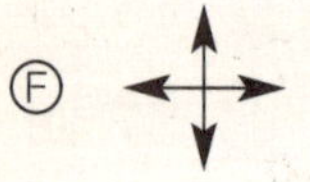

Ⓗ 

Ⓖ 

Ⓙ

❸ **Susie made the letter T using seven squares. Each square has a side equal to 2 cm in length. What is the perimeter of the T she made?**

Ⓐ 16 cm　　Ⓒ 28 cm

Ⓑ 14 cm　　Ⓓ 32 cm

Use a drawing to prove your answer.

Name ______________________________

1

$$\begin{array}{r} 22 \\ \times\ 24 \\ \hline \end{array}$$

Ⓐ 518

Ⓑ 132

Ⓒ 628

Ⓓ 528

Ⓔ none of these

2 **At the school district picnic, there were 308 fourth graders, 398 fifth graders, and 403 sixth graders. About how many students were at the district picnic?**

Ⓕ 1100 students

Ⓖ 1000 students

Ⓗ 900 students

Ⓙ 950 students

3

$$\begin{array}{r} 7.5 \\ -\ 2.3 \\ \hline \end{array}$$

Ⓐ 9.8

Ⓑ 5.2

Ⓒ 4.2

Ⓓ 5.8

Ⓔ none of these

4 $8 \times (\_\_\_\_ \times 6) = (8 \times 9) \times 6$

Ⓕ 48

Ⓖ 14

Ⓗ 72

Ⓙ 9

Name ____________________

1 **Which is the cylinder?**

 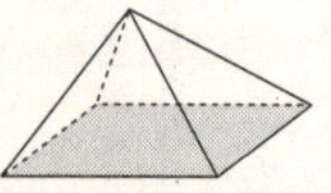

Ⓒ 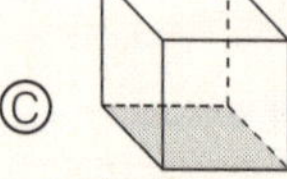

 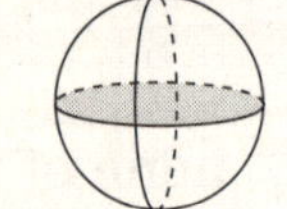

Ⓓ 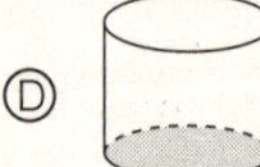

2

$$\frac{7}{10} + \frac{2}{10} =$$

Ⓕ $\frac{9}{20}$

Ⓗ $\frac{5}{10}$

Ⓖ $\frac{9}{10}$

Ⓙ $\frac{1}{2}$

3

$$3\overline{)153}$$

Ⓐ 501

Ⓒ 50 R1

Ⓑ 51

Ⓓ 44 R1

4 **Rearrange the digits 2, 4, and 7 to make as many different three-digit numbers as you can.**

Name ______________________________

1

$$\begin{array}{r} 6.9 \\ -\ 2.7 \\ \hline \end{array}$$

Ⓐ 9.6
Ⓑ 4.2
Ⓒ 4.3
Ⓓ 3.2

2 **How many months are in 10 years?**

Ⓕ 100 months
Ⓗ 50 months
Ⓖ 120 months
Ⓙ 80 months

3

$$\begin{array}{r} \frac{8}{10} \\ -\ \frac{5}{10} \\ \hline \end{array}$$

Ⓐ $\frac{1}{5}$
Ⓒ $\frac{1}{3}$
Ⓑ $\frac{3}{10}$
Ⓓ $\frac{3}{20}$

4

$$\begin{array}{r} \$1.50 \\ \times\quad 5 \\ \hline \end{array}$$

Ⓕ $7.50
Ⓖ $7.55
Ⓗ $5.50
Ⓙ $7.60

Name ______________________________

1

$$\begin{array}{r} 1065 \\ \times \quad 5 \\ \hline \end{array}$$

Ⓐ 50,028

Ⓑ 5,325

Ⓒ 5,532

Ⓓ 5,305

2

$$\begin{array}{r} \$\,21.83 \\ -\ 10.92 \\ \hline \end{array}$$

Ⓕ $11.91

Ⓖ $10.11

Ⓗ $11.11

Ⓙ $10.91

3 **406 is another name for:**

Ⓐ 4 hundreds 0 tens 6 ones

Ⓑ 4 hundreds 6 tens 0 ones

Ⓒ 4 tens 6 ones

Ⓓ 4 thousands 0 tens 6 ones

4

$$\begin{array}{r} \frac{6}{9} \\ +\ \frac{1}{9} \\ \hline \end{array}$$

Ⓕ $\frac{5}{0}$

Ⓗ $\frac{6}{81}$

Ⓖ $\frac{7}{9}$

Ⓙ $\frac{2}{3}$

Name ______________________________

1 **Which fraction below has the least value?**

Ⓐ $\frac{3}{8}$ Ⓒ $\frac{3}{4}$

Ⓑ $\frac{3}{5}$ Ⓓ $\frac{1}{2}$

3 **On one spin of the spinner, the probability it will stop on D is:**

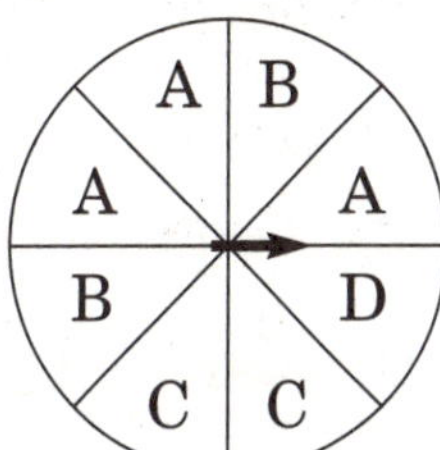

Ⓐ $\frac{1}{6}$ Ⓒ $\frac{1}{8}$

Ⓑ $\frac{2}{6}$ Ⓓ $\frac{3}{8}$

2 **How many factors does 10 have?**

Ⓕ 3 Ⓗ 4

Ⓖ 6 Ⓙ 5

4 **How many minutes until 5:00?**

Ⓕ 20 minutes

Ⓖ 35 minutes

Ⓗ 50 minutes

Ⓙ 25 minutes

Name ______________________________

1 **There are 8 soccer teams in the league. There are 12 players on each team. How many players are in the league?**

Select the correct number sentence.

Ⓐ $12 - 8 = 4$ Ⓒ $12 + 8 = 20$

Ⓑ $8 \times 12 = 96$ Ⓓ $8 + 12 = 20$

2

$$2\overline{)144}$$

Ⓕ 82 Ⓗ 702

Ⓖ 72 Ⓙ 62

3

$$\frac{1}{9} + \frac{1}{9} =$$

Ⓐ $\frac{1}{3}$ Ⓒ $\frac{2}{18}$

Ⓑ $\frac{2}{9}$ Ⓓ $\frac{1}{9}$

4 **Three multiples of 8 are:**

Ⓕ 8, 12, 20 Ⓗ 8, 10, 12

Ⓖ 1, 2, 4 Ⓙ 8, 16, 24

Name ______________________________

1 **Which drawing below shows perpendicular lines?**

Ⓐ 

Ⓒ 

Ⓑ 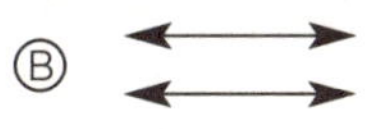

Ⓓ

2

$$\begin{array}{r} 4.4 \\ -\ 3.2 \\ \hline \end{array}$$

Ⓕ 1.2

Ⓖ 7.6

Ⓗ .2

Ⓙ 2.6

Ⓚ none of these

3

$$\begin{array}{r} 10 \\ \times\ 24 \\ \hline \end{array}$$

Ⓐ 240

Ⓑ 420

Ⓒ 204

Ⓓ 402

Ⓔ none of these

4 **Mr. Garcia had $20. He spent $13.82 at the market and $2.59 at the drug store. How much money does he have left?**

Ⓕ $17.41

Ⓗ $6.18

Ⓖ $3.59

Ⓙ $16.41

Name ____________________

1 **Late Monday night, a fisherman's boat ran out of gas 30 miles from shore. Early the next morning he began to row the boat toward shore. During the day he could row ten miles. Each night the current carried him back 2 miles. On what day of the week did he reach shore?**

____

Explain your answer with words, numbers, or a picture.

Name ______________________________

$4\overline{)164}$

Ⓐ 41

Ⓑ 401

Ⓒ 40 R1

Ⓓ 34 R1

❷

$$\begin{array}{r} 22 \\ \times\ 14 \\ \hline \end{array}$$

Ⓕ 110

Ⓖ 308

Ⓗ 36

Ⓙ 318

❸

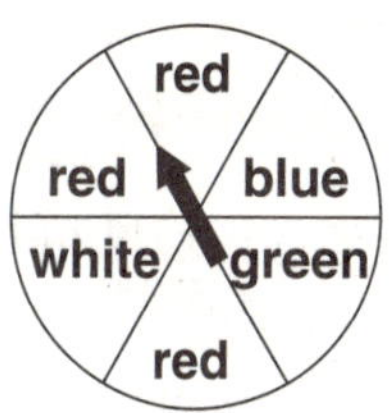

On one spin of the spinner, the probability it will stop on the red is:

Ⓐ $\frac{3}{6}$

Ⓑ $\frac{1}{6}$

Ⓒ $\frac{2}{6}$

Ⓓ $\frac{1}{3}$

❹

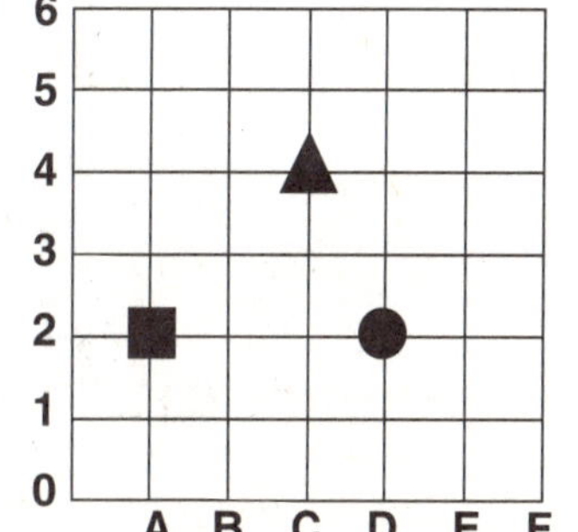

What is the point named by the ordered pair (D, 2)?

Ⓕ ●

Ⓖ ■

Ⓗ ▲

Name ______________________________

1 $4000 + 300 + 30 + 3 =$

Ⓐ 4,033　Ⓒ 40,333
Ⓑ 40,003　Ⓓ 4,333

2 $5\overline{)156}$

Ⓕ 31 R1　Ⓗ 32
Ⓖ 301 R1　Ⓙ 51 R1

3 **The seventh month of the year is:**

Ⓐ June　Ⓒ September
Ⓑ July　Ⓓ August

4 $\frac{2}{7} + \frac{3}{7} =$

Ⓕ $\frac{5}{14}$　Ⓗ $\frac{5}{7}$
Ⓖ $\frac{1}{7}$　Ⓙ $\frac{6}{49}$

1 2 3 4 5 6

INCH RULER **BELLWORK®**

CENTIMETER RULER **BELLWORK®**

1 2 3 4 5 6 7 8 9 10 11 12 13 14 15

1 2 3 4 5 6

INCH RULER

BELLWORK®

CENTIMETER RULER

BELLWORK®

1 2 3 4 5 6 7 8 9 10 11 12 13 14 15

Page ______________

A

1. Ⓐ Ⓑ Ⓒ Ⓓ Ⓔ
2. Ⓕ Ⓖ Ⓗ Ⓙ Ⓚ
3. Ⓐ Ⓑ Ⓒ Ⓓ Ⓔ
4. Ⓕ Ⓖ Ⓗ Ⓙ Ⓚ

Page ______________

B

1. Ⓐ Ⓑ Ⓒ Ⓓ Ⓔ
2. Ⓕ Ⓖ Ⓗ Ⓙ Ⓚ
3. Ⓐ Ⓑ Ⓒ Ⓓ Ⓔ
4. Ⓕ Ⓖ Ⓗ Ⓙ Ⓚ

Page ______________

C

1. Ⓐ Ⓑ Ⓒ Ⓓ Ⓔ
2. Ⓕ Ⓖ Ⓗ Ⓙ Ⓚ
3. Ⓐ Ⓑ Ⓒ Ⓓ Ⓔ
4. Ⓕ Ⓖ Ⓗ Ⓙ Ⓚ

Page ______________

D

1. Ⓐ Ⓑ Ⓒ Ⓓ Ⓔ
2. Ⓕ Ⓖ Ⓗ Ⓙ Ⓚ
3. Ⓐ Ⓑ Ⓒ Ⓓ Ⓔ
4. Ⓕ Ⓖ Ⓗ Ⓙ Ⓚ

Page ______________

E

1. Ⓐ Ⓑ Ⓒ Ⓓ Ⓔ
2. Ⓕ Ⓖ Ⓗ Ⓙ Ⓚ
3. Ⓐ Ⓑ Ⓒ Ⓓ Ⓔ
4. Ⓕ Ⓖ Ⓗ Ⓙ Ⓚ

Page ______________

F

1. Ⓐ Ⓑ Ⓒ Ⓓ Ⓔ
2. Ⓕ Ⓖ Ⓗ Ⓙ Ⓚ
3. Ⓐ Ⓑ Ⓒ Ⓓ Ⓔ
4. Ⓕ Ⓖ Ⓗ Ⓙ Ⓚ

Page ______________

G

1. Ⓐ Ⓑ Ⓒ Ⓓ Ⓔ
2. Ⓕ Ⓖ Ⓗ Ⓙ Ⓚ
3. Ⓐ Ⓑ Ⓒ Ⓓ Ⓔ
4. Ⓕ Ⓖ Ⓗ Ⓙ Ⓚ

Page ______________

H

1. Ⓐ Ⓑ Ⓒ Ⓓ Ⓔ
2. Ⓕ Ⓖ Ⓗ Ⓙ Ⓚ
3. Ⓐ Ⓑ Ⓒ Ⓓ Ⓔ
4. Ⓕ Ⓖ Ⓗ Ⓙ Ⓚ

Page ______________

I

1. Ⓐ Ⓑ Ⓒ Ⓓ Ⓔ
2. Ⓕ Ⓖ Ⓗ Ⓙ Ⓚ
3. Ⓐ Ⓑ Ⓒ Ⓓ Ⓔ
4. Ⓕ Ⓖ Ⓗ Ⓙ Ⓚ

Page ______________

J

1. Ⓐ Ⓑ Ⓒ Ⓓ Ⓔ
2. Ⓕ Ⓖ Ⓗ Ⓙ Ⓚ
3. Ⓐ Ⓑ Ⓒ Ⓓ Ⓔ
4. Ⓕ Ⓖ Ⓗ Ⓙ Ⓚ

Page ______________

K

1. Ⓐ Ⓑ Ⓒ Ⓓ Ⓔ
2. Ⓕ Ⓖ Ⓗ Ⓙ Ⓚ
3. Ⓐ Ⓑ Ⓒ Ⓓ Ⓔ
4. Ⓕ Ⓖ Ⓗ Ⓙ Ⓚ

Page ______________

L

1. Ⓐ Ⓑ Ⓒ Ⓓ Ⓔ
2. Ⓕ Ⓖ Ⓗ Ⓙ Ⓚ
3. Ⓐ Ⓑ Ⓒ Ⓓ Ⓔ
4. Ⓕ Ⓖ Ⓗ Ⓙ Ⓚ

Page ____________

A

1. Ⓐ Ⓑ Ⓒ Ⓓ Ⓔ
2. Ⓕ Ⓖ Ⓗ Ⓙ Ⓚ
3. Ⓐ Ⓑ Ⓒ Ⓓ Ⓔ
4. Ⓕ Ⓖ Ⓗ Ⓙ Ⓚ

Page ____________

B

1. Ⓐ Ⓑ Ⓒ Ⓓ Ⓔ
2. Ⓕ Ⓖ Ⓗ Ⓙ Ⓚ
3. Ⓐ Ⓑ Ⓒ Ⓓ Ⓔ
4. Ⓕ Ⓖ Ⓗ Ⓙ Ⓚ

Page ____________

C

1. Ⓐ Ⓑ Ⓒ Ⓓ Ⓔ
2. Ⓕ Ⓖ Ⓗ Ⓙ Ⓚ
3. Ⓐ Ⓑ Ⓒ Ⓓ Ⓔ
4. Ⓕ Ⓖ Ⓗ Ⓙ Ⓚ

Page ____________

D

1. Ⓐ Ⓑ Ⓒ Ⓓ Ⓔ
2. Ⓕ Ⓖ Ⓗ Ⓙ Ⓚ
3. Ⓐ Ⓑ Ⓒ Ⓓ Ⓔ
4. Ⓕ Ⓖ Ⓗ Ⓙ Ⓚ

Page ____________

E

1. Ⓐ Ⓑ Ⓒ Ⓓ Ⓔ
2. Ⓕ Ⓖ Ⓗ Ⓙ Ⓚ
3. Ⓐ Ⓑ Ⓒ Ⓓ Ⓔ
4. Ⓕ Ⓖ Ⓗ Ⓙ Ⓚ

Page ____________

F

1. Ⓐ Ⓑ Ⓒ Ⓓ Ⓔ
2. Ⓕ Ⓖ Ⓗ Ⓙ Ⓚ
3. Ⓐ Ⓑ Ⓒ Ⓓ Ⓔ
4. Ⓕ Ⓖ Ⓗ Ⓙ Ⓚ

Page ____________

G

1. Ⓐ Ⓑ Ⓒ Ⓓ Ⓔ
2. Ⓕ Ⓖ Ⓗ Ⓙ Ⓚ
3. Ⓐ Ⓑ Ⓒ Ⓓ Ⓔ
4. Ⓕ Ⓖ Ⓗ Ⓙ Ⓚ

Page ____________

H

1. Ⓐ Ⓑ Ⓒ Ⓓ Ⓔ
2. Ⓕ Ⓖ Ⓗ Ⓙ Ⓚ
3. Ⓐ Ⓑ Ⓒ Ⓓ Ⓔ
4. Ⓕ Ⓖ Ⓗ Ⓙ Ⓚ

Page ____________

I

1. Ⓐ Ⓑ Ⓒ Ⓓ Ⓔ
2. Ⓕ Ⓖ Ⓗ Ⓙ Ⓚ
3. Ⓐ Ⓑ Ⓒ Ⓓ Ⓔ
4. Ⓕ Ⓖ Ⓗ Ⓙ Ⓚ

Page ____________

J

1. Ⓐ Ⓑ Ⓒ Ⓓ Ⓔ
2. Ⓕ Ⓖ Ⓗ Ⓙ Ⓚ
3. Ⓐ Ⓑ Ⓒ Ⓓ Ⓔ
4. Ⓕ Ⓖ Ⓗ Ⓙ Ⓚ

Page ____________

K

1. Ⓐ Ⓑ Ⓒ Ⓓ Ⓔ
2. Ⓕ Ⓖ Ⓗ Ⓙ Ⓚ
3. Ⓐ Ⓑ Ⓒ Ⓓ Ⓔ
4. Ⓕ Ⓖ Ⓗ Ⓙ Ⓚ

Page ____________

L

1. Ⓐ Ⓑ Ⓒ Ⓓ Ⓔ
2. Ⓕ Ⓖ Ⓗ Ⓙ Ⓚ
3. Ⓐ Ⓑ Ⓒ Ⓓ Ⓔ
4. Ⓕ Ⓖ Ⓗ Ⓙ Ⓚ